Wild Chimpanzees

Social Behavior of an Endangered Species

As our closest primate relatives, chimpanzees offer tantalizing clues about the behavior of early human ancestors. This book provides a rich and detailed portrait of chimpanzee social life in the wild, synthesizing hundreds of thousands of hours of research at seven long-term field sites.

Why are the social lives of males and females so different? Why do groups of males sometimes seek out and kill neighboring individuals? Do chimpanzees cooperate when they hunt monkeys? Is their vocal behavior like human speech? Are there different chimpanzee "cultures"? Addressing these questions and more, Adam Arcadi presents a fascinating introduction to the chimpanzee social universe and the challenges we face in trying to save this species from extinction. With extensive notes organized by field site and an appendix describing field methods, this book is indispensable for students, researchers, and anyone else interested in the remarkable and complex world of these intelligent apes.

ADAM CLARK ARCADI is Associate Professor in the Department of Anthropology at Cornell University, where he teaches courses on human evolution, primate behavior, and primate conservation. He has conducted field research on wild chimpanzees in Kibale National Park (Uganda) and collaborated with researchers at the Gombe Steam Research Center (Tanzania) and the Taï Chimpanzee Project (Ivory Coast).

Wild
Chimpanzees

Social Behavior of an Endangered Species

ADAM CLARK ARCADI
Cornell University, New York

CAMBRIDGE
UNIVERSITY PRESS

University Printing House, Cambridge CB2 8BS, United Kingdom

One Liberty Plaza, 20th Floor, New York, NY 10006, USA

477 Williamstown Road, Port Melbourne, VIC 3207, Australia

314–321, 3rd Floor, Plot 3, Splendor Forum, Jasola District Centre, New Delhi – 110025, India

79 Anson Road, #06–04/06, Singapore 079906

Cambridge University Press is part of the University of Cambridge

It furthers the University's mission by disseminating knowledge in the pursuit of education, learning, and research at the highest international levels of excellence.

www.cambridge.org
Information on this title: www.cambridge.org/9781107197176
DOI: 10.1017/9781108178303

First published 2018

Printed in the United Kingdom by TJ International Ltd. Padstow Cornwall

A catalogue record for this publication is available from the British Library.

Library of Congress Cataloging-in-Publication Data
Names: Arcadi, Adam Clark, author.
Title: Wild chimpanzees : social behaviour of an endangered species / Adam Clark Arcadi, Cornell University, New York.
Description: Cambridge, United Kingdom ; New York, NY : Cambridge University Press, 2018. | Includes bibliographical references and index.
Identifiers: LCCN 2017061458| ISBN 9781107197176 (hardback : alk. paper) | ISBN 9781316647561 (paperback : alk. paper)
Subjects: LCSH: Chimpanzees – Behavior. | Endangered species – Behavior – Research.
Classification: LCC QL737.P94 A73 2018 | DDC 599.885–dc23
LC record available at https://lccn.loc.gov/2017061458

ISBN 978-1-107-19717-6 Hardback
ISBN 978-1-316-64756-1 Paperback

To the founders of all the long-term field studies. It is thanks to their courage, dedication, and stamina that we know so much.

Contents

The color plates are found between pages 146 and 147.

Preface

When the renowned fossil hunter Louis Leakey arranged for Jane Goodall to begin her field study of wild chimpanzees in East Africa, he hoped that her observations would shed light on the evolution of human ancestors. In the nearly six decades since, chimpanzees have become the most studied nonhuman mammal species in the wild and our primary model for thinking about how the last common ancestor of chimpanzees and humans may have behaved.[1] The list of intriguing discoveries about these African apes is long and still growing: wild chimpanzees are prodigious tool users, they form and manipulate coalitions to achieve social status, they hunt in groups for large mammal prey, they occasionally kill their neighbors, and they exhibit persistent group-level differences that are reminiscent of human cultural variation, to name a few of their most notable features. Revelations such as these offer important clues about the early stages of an evolutionary voyage that would lead to modern *Homo sapiens*, the most intelligent and socially complex animal in the history of life.

How sobering, then, to contemplate that these intelligent and socially complex apes, our closest relatives in the animal kingdom, are poised for extinction in their natural habitat. Chimpanzees form large, mixed-sex groups, or "communities," that aggressively defend extensive feeding territories where they find the ripe fruits that constitute the mainstay of their diet. When they reach sexual maturity, females typically disperse from these groups to breed in neighboring communities. In order for genetically viable populations to thrive, therefore, chimpanzees require intact forest areas that encompass multiple contiguous territories between which females can move. Habitat loss and fragmentation inevitably have devastating effects on chimpanzee populations, reducing food supplies, constricting mating opportunities, and forcing lethally antagonistic groups either into close contact with one another or into fringe habitats where they have access to only remnant patches of forest. Since the equatorial African rain forests on which chimpanzees rely continue to be harvested for timber and cleared for agriculture, their survival prospects are truly grim.[2]

Long-term research projects play a key role in chimpanzee conservation efforts. Protecting chimpanzee populations relies partly on having detailed and comprehensive information about their social behavior, their ecological requirements, and the limits of their social and ecological flexibility. Since chimpanzees are long-lived animals and tropical forest habitats vary greatly and change over time, this necessitates the establishment and maintenance of decades-long field studies. These studies involve numerous researchers, students, and field assistants monitoring known individuals across as many generations and in as many habitats as possible. Long-term projects also include natural scientists studying a wide range of topics in forest and community ecology, and sometimes social scientists exploring aspects of animal–human conflict, land history, and government policy. This interdisciplinary collaboration is essential for establishing successful forest preservation and management guidelines and for developing policies and practices that promote the coexistence of chimpanzees and the human neighbors with whom they periodically come into contact.

Long-term field projects also provide key conservation services beyond generating behavioral and ecological data.[3] The sustained presence of research staff in a forest deters encroachment and poaching by people from surrounding areas. Successful projects can employ dozens of local workers, providing an economic alternative to extracting forest resources. At the same time, researchers have the opportunity to collaborate with local people and, ideally, arrive at a mutual understanding of conservation goals and practices. Local field assistants and staff often develop more sympathetic attitudes toward forests and wildlife and share these attitudes with their communities. Researchers can educate themselves about community needs and historical land use patterns, an understanding of which can be critical to developing sustainable conservation schemes. Finally, scientific findings generate descriptions of chimpanzee social life that capture public attention, inspire students to pursue research, and attract donor support and tourist dollars. All in all, long-term field projects now constitute a bedrock component of chimpanzee conservation work.

This book provides an up-to-date synthesis of research on chimpanzee social behavior documented primarily from the seven field projects where it has been possible to monitor known individuals the most continuously for the longest time. Studies at these locations have been ongoing for more than twenty-five years; at two, research has been conducted with few interruptions since the 1960s, and at a few sites it has been possible at times to monitor neighboring communities simultaneously. These projects have collectively produced a rich picture of chimpanzee social life in the wild, based on hundreds of thousands of hours of observation by hundreds of researchers, and they continue to generate a steady flow of new data and surprises on this extraordinary species. There are also several other sites where wild chimpanzees have been fully habituated to the presence of researchers and can be followed at close range throughout the day. Relevant results from these projects are mentioned sporadically throughout, particularly when they reveal behavioral and technological diversity.

My goal is to provide a concise overview of wild chimpanzee behavior that is accessible to specialists and nonspecialists alike. I have therefore elected to keep the body of the text comparatively brief and to put citations, data, and supplementary information in endnotes. The number of scientific publications on chimpanzees is vast, and it is often difficult to keep track of which results come from which sites, what behaviors have been documented in all populations, and what questions remain to be examined in which communities. Therefore, where multiple studies on a specific topic are referenced, I list the studies in the endnotes by field site and always in the order that the long-term projects were initially established, i.e., Gombe studies first, then Mahale studies, and so on. In this way, it is immediately clear where the most work has been done and where information is lacking. For those interested in more detailed descriptions of particular chimpanzee communities, excellent volumes are available that document research on specific populations.[4] Likewise, readers can consult additional sources for descriptions of a second species of chimpanzee, the bonobo (*Pan pansicus*).[5]

I devote comparatively little attention to studies of captive chimpanzees, the literature for which is also vast and would require a book-length manuscript of its own to summarize. Behavioral research on captive chimpanzees, including widely publicized efforts to teach them rudimentary aspects of language, has contributed greatly to our understanding of chimpanzee cognition.[6] In addition, with the establishment of large groups housed in naturalistic conditions, researchers have been able to explore key aspects of development, the formation and maintenance of social relationships, and coalitionary behavior.[7] Nevertheless, although studies in captivity have the advantage of affording reliable observation conditions, their utility for exploring the evolution of species-typical behavior patterns is handicapped by the fact that captive animals are inevitably influenced by human contact and rearing conditions. Although at times I draw on insights gained from the study of artificially housed chimpanzees, for captivity often reveals the enormous behavioral potential of animals,[8] my focus is on behavior under natural conditions.

The book begins with an overview of the primates, situating chimpanzees within the order and describing the behaviors that researchers define and measure to quantify primate social behavior. This is followed in Chapter 2 by brief descriptions of the seven long-term field studies that have generated the most detailed information about chimpanzee social behavior in the wild. Nine subsequent chapters summarize well-researched areas of behavior, together constituting a coherent picture of chimpanzee social life in the wild across much of the species range. Results of studies on each behavior type are presented for each field site, to reveal both species-specific trends as well as inter-site variability. The Epilogue offers some final thoughts on chimpanzee conservation and the relevance of long-term field research. Lastly, the Appendix provides a brief description of field methodology for readers unfamiliar with animal behavior research in general and chimpanzee field research in particular.

This book would not have been possible without the support of many people. Meredith Small offered early advice on writing for a diverse readership and provided helpful comments on a draft of the Appendix. Chad Novelli, an avid nonfiction reader, was the first person to read a nearly complete draft of the manuscript and encouraged me to submit it for publication. Four anonymous readers offered helpful comments on the chapters submitted for review to Cambridge University Press. Martin Muller provided insightful comments on the completed manuscript, and Julia Fisher did likewise for Chapter 10. Richard Wrangham generously provided access to the Kibale Chimpanzee Project photo collection, helped acquire permissions, and answered many questions about the Kanyawara chimpanzees. Aggrey Rwetsiba, Senior Monitoring and Research Coordinator for the Uganda Wildlife Authority, kindly granted permission to include the color photographs, which were taken for research purposes. Andrew Bernard, Ronan Donovan, and John Mitani kindly provided wonderful photographs. At Cambridge University Press, Megan Keirnan and Noah Tate provided essential help and advice throughout the publication process.

Finally, teaching keeps me in touch with the fascination that drew me to the study of chimpanzees, and reminds me constantly of how transporting it is to track and observe them in the wild. With the pressure to publish scientific papers on exceedingly narrow topics, it is easy to forget the quiet sense of wonder that permeates even the most rigorous days collecting field data. I am indebted to my undergraduate students for inspiring me to attempt a synthesis of wild chimpanzee behaviorial research that would offer readers the opportunity to enter into the astonishingly complex social world of these intelligent apes and to reflect on the differences that make humans so unique.

1

Primates, Apes, and the Study of Chimpanzee Social Behavior

The mammalian order Primates includes roughly 500 species, most of which are found in the tropics.[1] As a group, primates are characterized by a constellation of traits that reflect adaptations to arboreal habitats, dietary flexibility, and extensive parental care. Not every trait is present in every species, and when they are present, most traits are expressed to varying degrees in different groups. Nevertheless, all primates possess most or all of the following:[2]

- Flexible and dexterous hands and feet, with flattened nails and sensitive tactile pads on the tips. All species except humans have opposable big toes, and most have opposable thumbs, permitting enhanced grasping ability with both the feet and hands.
- A generalized dentition, typically with four types of teeth, supporting the ability to process a wide range of foods.
- Excellent vision, including the ability to see color, which is absent in other placental mammals.[3]
- Comparatively large brains for their body size, with a longer period of infant and adolescent development, accompanied by an increased role for learning in acquiring social and technical skills.

The order is divided into two suborders, the Strepsirrhini and the Haplorrhini (Figure 1.1).[4] The strepsirrhines include the more "primitive" species, meaning those that appear to resemble most closely the earliest primate ancestors in body plan and, presumably, behavior.[5] Strepsirrhines have a well-developed sense of smell, eyes that are not completely forward facing and that lack a protective bony socket, and a characteristic form of locomotion called "vertical clinging and leaping."[6] A large percentage of strepsirrhines are small and nocturnal and do not form permanent social groups. Like many other mammals,

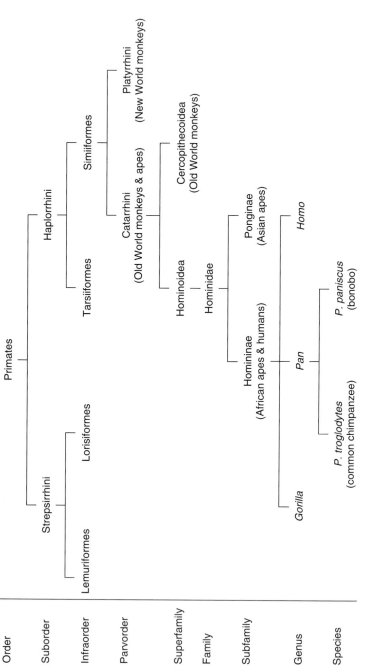

Fig. 1.1 Abbreviated taxonomy of the Primate order, showing the location of chimpanzees within the hominoid superfamily (following Groves, 2001).

most strepsirrhines are seasonal breeders, with females remaining sexually receptive for narrowly limited periods of time each year.[7]

The haplorrhines include the monkeys, the apes, and humans. With the exception of the several owl monkey species (genus *Aotus*) in Central and South America, haplorrhines are diurnal. They rely greatly on sight, which is reflected in several morphological traits: Their eyes are more forward facing, enhancing depth perception (stereopsis), and are housed in bony sockets.[8] Additionally, their retinas are more sensitive, and they have a comparatively large area of brain cortex associated with visual processing.[9] Consistent with their greater reliance on sight, visual signaling plays an important role in haplorrhine social interactions. The majority of haplorrhines form permanent social groups with multiple adult females and at least one permanent male member. They also have relatively larger brains than strepsirrhines, associated in general with more complex social behavior.

THE APES (SUPERFAMILY HOMINOIDEA)

At the peak of their ecological success, during the Miocene epoch, scores of ape species existed throughout Asia, Europe, and Africa.[10] As a consequence of the gradual reduction in tropical forest cover caused by cooler temperatures and changing rainfall patterns worldwide, most of these apes went extinct by the end of the Miocene. Today only a handful of species remain, confined to dwindling tropical forests in southeast Asia and Africa. The Asian apes include twelve to fourteen species of gibbons (family Hylobatidae, the small-bodied "lesser apes") and two species of orangutans (*Pongo pygmaeus* and *P. abelii*, members of the "great apes," restricted to the islands of Borneo and Sumatra, respectively).[11] Four species of African great apes also survive: two species of gorillas (*Gorilla gorilla*, western gorillas, and *G. beringei*, eastern gorillas) and two species of chimpanzees (*Pan troglodytes*, the common chimpanzee, and *P. paniscus*, the bonobo).[12]

Current fossil and genetic evidence indicates that the last common ancestor of gibbons and the great apes lived roughly 16 to 18 million years ago (mya).[13] Among the great apes, the orangutans shared a common ancestor with the African apes 12 to 16 mya. The gorilla and chimpanzee lineages appear to have split 9 to 12 mya, while the last common ancestor of humans and the two chimpanzee species lived 6 to 9 mya. Common chimpanzees (hereafter referred to simply as "chimpanzees") and bonobos diverged after the split from the human lineage, around 1 mya.[14] The fossil record for the African apes is poor, and consequently it is unclear how much their lineages changed in appearance and behavior since these branching points. By contrast, many bipedal species evolved and disappeared over the past 5 million years, first in Africa and subsequently in Europe and Asia. Some fraction of these bipedal taxa were direct ancestors of modern humans.[15]

A number of general features distinguish the apes (and humans) from other primates. They lack tails as well as facial hair, the latter enhancing facial expressiveness and visual communication. Ape shoulder joints allow full rotation of the arms, an adaptation that enables them to hang or swing efficiently under branches – whereas monkeys primarily walk along the top surfaces of canopy branches to travel, apes frequently hang beneath branches and move arm over arm in a form of locomotion referred to as "brachiation" or "suspensory locomotion" (humans can employ a form of suspensory locomotion on "monkey bars," the horizontal ladder that is a standard feature on playground equipment). Apes have comparatively large brains and extended periods of development during which infants and juveniles acquire social and technical skills. In addition, their larger comparative brain size, and more specifically their enlarged frontal cortex, is associated with enhanced cognitive complexity compared with monkeys.[16]

The different ape genera also exhibit distinctive locomotor specializations. Gibbons are quintessential brachiators. They have comparatively long arms and fingers, short legs, and reduced thumbs, and with their small body size can swing exceptionally fast and acrobatically through the forest canopy. They commonly walk bipedally when they are on the ground. The large-bodied apes usually move more slowly through the canopy and generally walk quadrupedally on the ground to travel. Orangutans walk on the soles of their feet and on the backs of their hands, which are balled into fists (called "fist-walking"). The African apes also walk on the soles of their feet, but on the knuckles of their partly flexed hands ("knuckle-walking"; specifically, on the back surface of the middle phalanges or finger bones). Humans, of course, walk and run bipedally. Although with significant training we can become competent at climbing trees, our lack of an opposable big toe makes it impossible to attain the skill of an ape.[17]

Ape Diets and Daily Activity Patterns

Ripe fruits predominate in the diets of all apes, with the exception of mountain gorillas, for whom fruits are less readily available (Plate 1).[18]

Leaves, flowers, seeds, pith, and even bark from a multitude of trees, shrubs, and herbaceous plants are also utilized – some populations of chimpanzees feed on more than 200 different plant species, although they rely on some types more than others.[19] Variability between populations within ape species in the percentage of the overall diet represented by fruits and other plant parts is linked to botanical diversity and productivity in different forest areas.[20] All apes consume insects to varying degrees, especially ants and termites. Meat eating is generally rare, except in chimpanzees, who are avid hunters and prey on a wide variety of mammal species (Chapter 9).

The timing of fruit production in tropical forests is variable, both within and between tree species.[21] Apes must therefore search continuously for ripe fruits, covering distances each day ranging from a few hundred meters to many kilometers (Chapter 4).[22] The area utilized by individuals over the course of a year, their "home range," varies widely depending on their nutritional requirements, habitat characteristics, and competition with group members and neighbors. Home ranges vary from as little as 0.2 km^2 in some gibbon groups to 50.0 km^2 for chimpanzees living in arid habitats. In gibbons and chimpanzees, a portion of the home range, referred to as a group's "territory," is aggressively defended (Chapter 8).

Bouts of travel and feeding that begin early in the morning are interspersed with periods of resting and, in group contexts, socializing on the ground or in the canopy. Members of all species end the day by choosing a new sleeping site. Gibbons, orangutans, and chimpanzees sleep high in trees, whereas gorillas sleep either low in trees or on the ground.[23] Adolescent and adult great apes build a new sleeping "nest" each evening prior to nightfall by bending and weaving branches together into a rough platform that can support their weight. Occasionally individuals also construct more makeshift nests to rest in during the day. Nests are rarely reused, and individuals do not share them to sleep together, with the exception of dependents and their mothers.[24] Individuals may wake and produce vocalizations from their nests but they rarely travel to new locations during the night.[25]

Ape Social Systems and the Mother–Infant Bond

The "social system" of a primate species has three quantifiable components.[26] First, "social organization" refers to the size, sexual composition, and cohesion of the groups in which individuals regularly associate. Second, patterns of individual social relationships, both within and between the sexes, represent the "social structure" of the group. Finally, the "mating system" refers to how many members of the opposite sex individuals typically copulate with. Four basic mating systems are recognized in primates: monogamy (one male and one female mate only with each other over the course of many years), polygyny (one male mates with many females, while the females only mate with that male), promiscuity (both males and females mate with multiple members of the opposite sex), and polyandry (one female mates with more than one male and the males only mate with that female).

The apes exhibit a range of social systems, and as with other primates, typical patterns can vary within species. Gibbons usually live in groups with one adult male, one adult female, and younger individuals, but some groups may temporarily include additional adults of either sex.[27] Although historically classified as monogamous, both male and female gibbons may seek matings with individuals in other groups, and both polygynous and polyandrous groups are occasionally observed. Orangutan social organization is referred to as "solitary," since adults do not associate in cohesive social groups, although a form of dispersed group structure may exist in some populations.[28] Dominant males mate polygynously, attempting to defend exclusive sexual access to several females whose home ranges lie within their own, while subordinate males roam widely. Gorilla groups contain several females and one or, less often, multiple males; temporary all-male groups also form, comprised of individuals who have not succeeded in forming a group with females.[29] Like dominant orangutan males, dominant gorilla males mate polygynously as a rule. Finally, chimpanzees and bonobos live in multi-female, multi-male groups ranging in size from 20 to roughly 200 individuals, and their mating system is promiscuous (Chapter 7).

Significant differences as well as similarities exist between the apes in patterns of social relationships within groups. The closest adult relationships in gibbons and gorillas, taxa in which both sexes typically disperse to breed, occur between males and females. In gibbon groups that contain an additional adult male or female, individuals of the same sex may be mutually tolerant and somewhat affiliative.[30] The adult females in gorilla groups are usually unrelated and are generally tolerant but indifferent to each other, focusing their social attention instead on the dominant male. This male provides protection from outsider males and predators, and intervenes in conflicts between females competing for resources. In multi-male gorilla groups, which often include sons that have not dispersed, male relationships are generally tense.[31] By contrast, the strongest adult bonds are formed among males in chimpanzees (Chapter 6) and among females in bonobos. In both of these species, it is females that disperse to breed in neighboring groups (Chapter 3).[32]

Yet despite these differences in adult patterns, the most consistently strong relationships in all ape species are between mother and offspring. Periods of infant and juvenile dependency lasting many years are characterized by high levels of physical proximity and contact, grooming, protection, and play (Chapter 5 and later in this chapter). The intensity and importance of the mother–offspring bond in chimpanzees, for example, is revealed when one of them dies: Mothers may carry a dead infant for weeks before finally relinquishing it, while even weaned juveniles may fail to thrive after the death of their mother.[33] Among chimpanzees and bonobos, species in which males remain for life in their natal group, enduring relationships between mothers and sons are common.[34] Analogous father–son relationships do not exist, since females mate with many males and fathers apparently cannot recognize their offspring (Chapter 7).

Triadic Awareness

The ability to maintain a variety of social relationships with known individuals within stable groups is characteristic of, but not unique to, group-living primates.[35] What is perhaps especially well developed among primates, however, is the ability to recognize the varied relationships of other individuals, and to interact with those individuals accordingly. For example, immediately following aggressive encounters with group mates, fight participants sometimes direct aggression at a third, uninvolved individual, a behavior known as "redirected aggression." In many monkey species in which females remain in the groups they are born into and form close, lifelong relationships with their female relatives, redirected aggression is often targeted at relatives of the aggressor's original opponent.[36] Male mountain gorillas intervene in conflicts between females without taking sides, perhaps in effect maintaining a protective role toward all and discouraging females from transferring out of their breeding groups.[37] And male chimpanzees are keenly aware of the relative dominance ranks of other males as they vie for coalition partners during dominance conflicts (Chapter 6).

In at least some monkey species, this so-called "triadic awareness"[38] appears to involve the ability to simultaneously classify group members according to both rank and membership in matrilines. It also includes classifying and tracking important temporary associations, such as between a male and an estrous female.[39] Whether chimpanzees and other social species characterized by "fission–fusion" social organization (see later) may have an even richer capacity for triadic awareness is unknown.[40] In any case, observations both in the wild and under experimental conditions suggest that increased comparative brain size and, by extension, intelligence in primates evolved because of the advantages individuals gained by being better able to navigate complex social landscapes.[41]

QUANTIFYING CHIMPANZEE SOCIAL BEHAVIOR TO ASSESS INDIVIDUAL RELATIONSHIPS

Social relationships are formed through the accumulation over time of behavioral interactions.[42] Researchers characterize social relationships by quantifying the relative frequency, duration, and intensity of "affiliative," "cooperative," and "agonistic" behaviors that individuals direct toward each other (roughly speaking, behaviors that are friendly, supportive, or associated with aggression and submission).[43] Individuals that interact peacefully with each other at especially high rates are described as being "socially bonded."[44] Socially bonded individuals, in turn, are often more likely to engage in mutually supportive behaviors. By contrast, agonistic interactions reflect conflicts of interest between individuals. Agonistic encounters form the basis of dominance relationships, which in turn generally determine access to contested resources. As described previously, patterns of individual social relationships then constitute the social structure of a species, which together with social organization and mating system comprise the species' social system (or "society").

Researchers typically begin field studies by observing, describing, and classifying social and other behaviors exhibited by their study animals. The resultant behavioral catalog is known as an "ethogram." Because primate social behaviors frequently seem familiar to us, the risk of attributing human motivations and intentions to them is high. Ethogram entries thus tend to be rather sterile descriptions of physical movements, in an effort to avoid anthropomorphic bias and improve consistency in data collection within and between study sites. For socially complex species such as chimpanzees, ethograms grow over time as new behaviors are observed, resulting ultimately in extensive behavioral records.[45] A relatively narrow group of behaviors within the ethogram are used to build a statistical picture of social relationships. Some of the key behaviors used to characterize chimpanzee relationships are described in the following sections.

Affiliative Behaviors

Spending Time Together

Chimpanzees forage, socialize, and rest in small, spatially separate subgroups within their community throughout the day. Membership in these subgroups changes over time, so that individuals may be in the company of different community members on different days or even at different times on the same day, a grouping pattern referred to as fission–fusion social organization (Chapter 3).[46] The amount of time two individuals spend near each other consequently represents an initial indicator of social bonding between them, on the premise that friendly individuals seek out each other's company. The Dyadic Association Index (DAI), which quantifies the proportion of total observation time that pairs of individuals are seen together, characterizes this aspect of relationships (Chapter 4, endnote 15). Additionally, DAIs are used to compare patterns of social attraction between different classes of individuals within chimpanzee communities.[47]

Measures of association, however, are only suggestive of social affinity, since mutual attraction to food sources or other community members can also bring individuals into proximity. To develop a more accurate picture of social relationships, researchers therefore rely on analyses of the frequency and duration of social grooming and the incidence of cooperative behaviors. These behaviors are overtly directed at particular individuals and can involve a significant commitment of time and energy, therefore implying a higher degree of social bonding.

Social Grooming

Social grooming is a tactile and highly visual behavior in which one individual closely scrutinizes and meticulously picks through the hair of another, removing debris and insects with their hands and/or mouth.[48] It is a pervasive feature of chimpanzee life and occurs in a wide range of contexts. The subjective impression reported by human observers is that grooming is pleasurable to chimpanzees, insofar as individuals being groomed appear relaxed and may allow groomers to manipulate their body parts in the process. Grooming sessions can last anywhere from a few minutes to two hours or more, representing a significant commitment of time that might otherwise be spent on feeding or resting.[49] Two isolated individuals may take turns grooming each other or groom each other simultaneously (referred to as "mutual grooming"). Groups of chimpanzees, sometimes ten or more, may groom in close proximity, often with several individuals grooming in a chain (A grooms B while B grooms C, etc.) or pairs of individuals grooming a third.

Results from studies in a variety of primate species, including chimpanzees, indicate that grooming is important socially. Individuals spend more time grooming in species that form larger groups, suggesting that it helps them navigate more complex social landscapes.[50] In contrast, individuals in large-bodied species do not groom more than those in small-bodied species, which would be expected if grooming merely served a hygienic function.[51] In some species, the likelihood of grooming increases when social relationships are unstable and after agonistic interactions, apparently helping to alleviate stress. This conclusion is supported by analyses showing that grooming reduces heart rates, circulating levels of glucocorticoid hormones, and behavioral indicators of stress (e.g., self-scratching).[52] Finally, grooming appears to play a key role in the formation and maintenance of social bonds, whether by simply promoting tolerance around more dominant and potentially aggressive individuals, or by strengthening cooperative relationships by reinforcing familiarity, trust, and predictability between partners.[53] The bond-maintenance effect of grooming is additionally supported by evidence that circulating levels of oxytocin, a hormone that elevates affiliative behavior in mammals, increase in socially bonded chimpanzees after they groom together.[54]

Social Play

Chimpanzees of all ages and both sexes play, although it is most common among juveniles and adolescents. Social play can take the form of a seemingly infinite variety of both vigorous and gentle behaviors, including chasing, grappling, tickling, "finger wrestling," sexual play, and play mothering.[55] Like social play in other animals, it is widely assumed to provide important opportunities to develop physical and social skills. However, since social play is rarely the focus of long-term research, no doubt in part because it is difficult to assess its impact on adult social life, few quantitative data are available to explore its precise functional significance.[56]

Cooperative Behaviors

Three prominent cooperative behaviors are observed regularly in wild chimpanzees. First, two or more individuals may join together to attack another chimpanzee. Such coalitions can occur in a variety of within-group and between-group contexts and involve either males or, albeit less frequently, females (Chapters 5, 6, and 8). Second, male chimpanzees silently patrol territorial boundary areas together, closely coordinating their movements and engaging neighboring individuals in unison (Chapter 8). Finally, groups of chimpanzees commonly hunt mammalian prey together and share meat after successful captures. Although it appears unlikely that individuals coordinate their actions to a significant degree when pursuing prey, the subsequent sharing of kills is more clearly cooperative. Importantly, these cooperative behaviors appear to be reciprocated, both in kind as well as between types – for example, some studies have shown that individuals that groom together regularly are more likely to support each other in aggressive interactions, engage in territorial behavior together,[57] and share meat following predatory events (Chapter 9).[58]

Agonistic Behaviors

Like other primates, chimpanzees exhibit a variety of behaviors associated with aggression and submission. Physical attacks can include hitting with hands and feet, grappling, twisting limbs, and biting. Dominance relationships are established based on the outcomes of physical confrontations and reduce the subsequent occurrence of risky aggression. Several types of vocalizations and physical gestures signal threats and submission, and appear to reduce the likelihood of chasing and fighting (Chapter 10).[59] Lower-ranking individuals indicate their subordinate status with a characteristic vocalization, the "pant grunt," typically move out of the way of higher ranking individuals who walk near them ("avoids"), and relinquish feeding locations to them upon approach ("supplants"). Since physical fights are comparatively rare, researchers typically use the frequency and direction of pant grunts, avoids, and supplants to quantify dominance relationships among their study animals.

MAINTAINING SOCIAL RELATIONSHIPS

Membership in a permanent social group is a necessity for chimpanzees – survival in isolation appears rare if not impossible (Chapter 8). The ability to form social relationships is therefore crucial for these animals. But with group existence also comes conflict, since individuals that rely on each other for protection and access to resources also compete with each other constantly for status, food, and mates. Consequently, behaviors that promote social bonding must be complemented with mechanisms to maintain positive social relationships in the face of stress and potentially damaging conflicts. This chapter concludes with a brief description of the behaviors chimpanzees exhibit that appear to safeguard vital social relationships, and thus underpin stable social groups.

Calming Physical Contact

Chimpanzees regularly engage in many types of apparently calming physical contacts, a few of which are noted here.[60] Individuals reuniting after periods of separation exhibit what researchers refer to as "greeting" behaviors. These include briefly touching hands or lips together (the latter referred to as "kissing"), or one individual placing a hand on the other's body or extending a hand to the other's lips. Moments of fear or social excitement elicit "reassurance" behaviors: Individuals may briefly embrace one another or touch each other's genitals. Particularly tense individuals signal submission to more dominant group members by approaching in a low, crouching posture, often bobbing head and body while reaching out with a hand (Chapter 10). These gestures may be responded to with a brief touch of the hand by the higher-ranking individual, apparently signaling tolerance and facilitating subsequent peaceful proximity between the two.

Repairing Social Relationships with Peaceful Post-Conflict Interactions

Soon after the initial physical separation that inevitably follows aggressive interactions between chimpanzees, aggressors and victims frequently engage in peaceful contact behaviors (patting, embracing, mouth-to-mouth touching) with each other and/or nearby group members. These reassurance behaviors were recognized and described in the earliest long-term field studies.[61] In subsequent analyses among captive chimpanzees and rhesus macaques, researchers developed a standardized post-conflict/matched control (PC/MC) methodology (see endnote) to confirm that such calm physical contacts were a direct response to the preceding aggressive incident.[62] Two basic patterns of peaceful post-conflict interactions (PPCI) are now distinguished. "Reconciliation" is defined as contact between attacker and victim, and it typically occurs within a few minutes after the interaction. "Consolation" is defined as contact between one of the fight participants and a third individual not involved in the actual fight.

Among captive chimpanzees, between 17 and 48 percent of aggressive conflicts are followed by reconciliation. Reconciliation rates among wild chimpanzees are generally lower, between 12 and 21 percent, perhaps because individuals can more easily distance themselves from opponents while allowing tension to dissipate.[63] Studies in a variety of primate species indicate that reconciliation has the immediate effect of reducing tension and facilitating tolerance soon after agonistic interactions.[64] Importantly, variation between pairs of individuals in the propensity to reconcile suggests the hypothesis that reconciliation functions to repair valuable relationships disrupted by conflict.[65] This idea has been broadly supported by the observation across many primate species that the likelihood of reconciliation increases when individuals are more strongly socially bonded. Among male wild chimpanzees, for example, reconciliation is more likely between individuals that groom together more and between individuals that form coalitions together (Chapter 6).[66]

The function of consolation, or approach and contact with a fight participant by a third party after a conflict, is less clear and may be more variable.[67] Since third-party contact sometimes appears to have a calming effect on both aggressors and victims, researchers initially concluded that it functioned to "console" fight participants, and some studies in captivity have found that third-party contact reduces stress and is more likely between individuals with close relationships.[68] Other studies in captivity, however, have failed to document this effect.[69] Instead, there is evidence that post-conflict bystander affiliation can reduce the likelihood that the bystander him- or herself will be a target of ongoing aggression (the "self-protection hypothesis").[70] Similarly, some evidence has been advanced supporting the hypothesis that post-conflict bystander affiliation might help opponents to reconcile ("opponent relationship repair hypothesis").[71] Although the ultimate function of third-party contact remains elusive, it is evident that the patterning and consequences of the behavior are strongly influenced by the nature of the relationships between the several interacting individuals.

Finally, although chimpanzees share with other primates the basic behavioral mechanisms for forming and maintaining social relationships, the patterning of those relationships differs in important ways from most other species. In the majority of catarrhine primates, it is either males or both males and females that leave the group that they are born into and join a new group to reproduce when they reach sexual maturity. As mentioned previously, in chimpanzees it is the females that typically emigrate from their natal group. As the following chapters will show, this has profound implications for social relationships both within and between the sexes.

2

Seven Long-Term Field Studies

Between 150,000 and 300,000 chimpanzees survive in the wild, living in mostly unprotected forested or partially forested habitats distributed discontinuously across equatorial Africa.[1] Interdisciplinary research is being conducted at a growing number of locations, primarily in areas enjoying some level of legal protection. Well-funded projects are able to staff field stations year-round to continuously collect ecological, demographic, and behavioral data and maintain digital databases to which project leaders, visiting scientists, students, and local field assistants alike contribute. The emergence of digital technologies has also facilitated collaborative analyses among different field projects, substantially enlarging our ability to explore the extent and significance of variation in chimpanzee behavior across their geographical range. Long-term projects generate critical information and support for conservation initiatives and can be instrumental in elevating and maintaining the protected status of the forests in which chimpanzees are located.

BRIEF FIELD SITE DESCRIPTIONS

The locations of the seven field studies that have generated the most detailed information about wild chimpanzee social behavior are shown in Figure 2.1. Each of these projects has amassed decades-long records of socioecological data for populations in which individual identities are known and genealogical relationships have been tracked for generations. The following sections provide brief overviews of the establishment, general scientific focus, habitat characteristics, and chimpanzee population status of each project. Most of these projects maintain websites at which full lists of published research may be found along with information about conservation and education

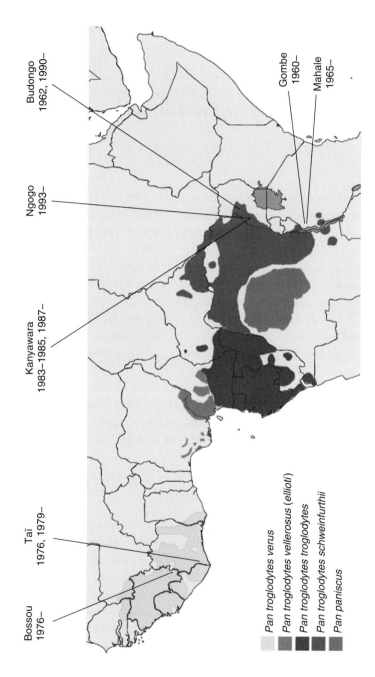

Fig. 2.1 Locations of the seven long-term field studies that have generated the most data on wild chimpanzee social behavior. See text for details. (A black-and-white version of this figure will appear in some formats. For the color version, please refer to the plate section.)

Bossou
1976–

Taï
1976, 1979–

Kanyawara
1983–1985, 1987–

Ngogo
1993–

Budongo
1962, 1990–

Gombe
1960–

Mahale
1965–

Pan troglodytes verus
Pan troglodytes vellerosus (ellioti)
Pan troglodytes troglodytes
Pan troglodytes schweinfurthii
Pan paniscus

activities. The field sites are presented in the order in which they were established chronologically.

Gombe Stream Research Center (Gombe National Park, Tanzania)

Jane Goodall initiated her pioneering study of chimpanzees in what is now Gombe National Park, on the eastern shore of Lake Tanganyika, in 1960. A series of televised National Geographic Society productions featuring her early research dramatized discoveries of close family relationships, tool use, and hunting, and brought chimpanzees into the public eye. Tanzanian field assistants and a steady stream of university researchers have been collecting information at Gombe ever since. More than five decades of data are now managed by the Jane Goodall Institute Research Center at Duke University, providing a critical resource for examining the ecological status of the Gombe chimpanzees and the evolutionary significance of a wide range of chimpanzee behaviors.

Gombe National Park covers an area of 35 km^2 (14 x 2–3.5 km, or 13.5 mi^2) and is comprised of a mixture of riverine forest, deciduous woodland, grassland, and thicket. There are two dry and two rainy seasons, with an overall average annual rainfall of approximately 1,500 mm. It is located on a cluster of steep hills divided by streams flowing into Lake Tanganyika. As of 2010, there were 101 to 105 chimpanzees at Gombe divided between three distinct communities and largely cut off from other populations by agricultural development. The "Kasekela" community, the primary study group, occupies an area that varies between 5 and 13 km^2 and has had between thirty-eight and sixty-two individuals since 1966. The "Kalande" and "Mitumba" communities, to the south and north, contained roughly fifteen to nineteen and twenty-five individuals, respectively, as of 2010.[2]

The Kasekela chimpanzees were provisioned daily with bananas from 1962 to 1967 to make them less fearful of humans and easier to observe. Levels of provisioning were reduced over the next two years, and between 1969 and 2000, individual chimpanzees received bananas every seven to ten days. Provisioning was finally terminated in 2000.[3] The Mitumba community was also provisioned from 1985 to 2000.[4] The research staff at the field station occasionally provides veterinary care to sick or injured animals. For example, polio vaccine and antibiotics have been administered via bananas, and old and sick individuals are sometimes offered foods.[5] In general, however, contact between humans and chimpanzees is now strongly discouraged since chimpanzees can and do contract human diseases, sometimes resulting in lethal epidemic outbreaks.[6]

Mahale Mountains Chimpanzee Research Project (Mahale Mountains National Park, Tanzania)

Toshisada Nishida initiated long-term research in 1965 in what is now the Mahale Mountains National Park. Nishida was subsequently the first to discover and describe the fission–fusion structure of chimpanzee social groups (Chapter 3).[7] He was also the first to discover that female chimpanzees emigrate permanently from their natal communities to breed,[8] in contrast to a common pattern in monkey species wherein females remain in the groups they are born into and associate preferentially with female relatives. As at Gombe, scientists and field assistants at Mahale continue to document chimpanzee natural history in tremendous detail and maintain a vast database for comparative and long-term analyses.[9]

The Mahale Mountains National Park covers an area of 1,613 km^2 (623 mi^2) and is home to an estimated 700 chimpanzees.[10] The chimpanzee study area is about 160 km (100 miles) south of Gombe and likewise located on the eastern shore of Lake Tanganyika. The habitat is a combination of riverine forest and semi-deciduous mid-altitude and montane rain forest (mean rainfall 1,500–2,500 mm/year). As at Gombe to the north, streams flowing into the lake separate steep, rugged, forested slopes spanning 800 to 1,300 meters above sea level (m.a.s.l.). The first group Nishida studied, "K-group," was comprised of thirty individuals occupying a range of about 6 km^2.[11] Nishida and his colleagues began study of a neighboring community to the south, known as "M-group," in 1968. In the early 1980s, all of K-group's males disappeared, and the females from this group subsequently dispersed (Chapter 9). Since then, M-group has been the primary study community. It currently includes about sixty individuals and occupies an area of roughly 25 km^2.[12]

As at Gombe, the Mahale chimpanzees were initially provisioned in order to make them easier to find and follow. K-group and M-group received sugarcane and bananas beginning in 1966 and 1968, respectively. Provisioning was finally abandoned in 1987.[13] Mahale researchers have occasionally intervened directly in the social life of the study animals. For example, in 1981 several field workers prevented a group of males from attacking a female and her infant when it seemed likely that the infant would be killed (see Chapter 9 for discussion of infanticide).[14]

Bossou-Nimba Chimpanzee Research Project (Republic of Guinea)

Chimpanzees were surveyed and studied briefly at Bossou, in southwestern Guinea, West Africa, by Dutch scientists in the 1960s.[15] Between 1976 and 1986, a period when the country's political climate made it difficult to sustain uninterrupted research, Yakimura Sugiyama conducted three four- to seven-month field seasons observing a small and isolated group of chimpanzees living in close proximity to human settlements.[16] In 1986, Tetsuro Matsuzawa joined Sugiyama, and together they established a permanent research presence at Bossou. Researchers at Bossou have focused their work especially on chimpanzee tool use and its social transmission. The Bossou study group is isolated from neighboring chimpanzee populations, and its unique demographic history, social organization, and coexistence with nearby human populations constitute an important case study of the potential of chimpanzees to adjust to ecological disturbance.

The Bossou chimpanzees were habituated without provisioning, although they have occasionally been offered bananas by researchers conducting experiments on tool-using behavior.[17] They utilize a roughly 6 km^2 area of forest covering a series of small, 70- to 150-meter-high hills surrounding the village of Bossou, and make periodic forays along wooded corridors to nearby forest patches. The habitat is comprised of a mixture of primary and regenerating secondary forest, situated within an agricultural region dotted with small villages and fields. During the long rainy season from March to October, monthly rainfall can reach 700 mm. The group maintained a stable population size of sixteen to twenty-two individuals until 2003, when the outbreak of a respiratory illness reduced its numbers to twelve. The community has been stable since (as of 2010). The nearest neighboring population is in the Nimba Mountains, 6–10 km away.

The study animals come into regular contact with nearby humans when crossing roads and paths that cut through forest patches and when raiding crops. The forests and chimpanzees are considered sacred by the local Manon people, and consequently villagers and chimpanzees have coexisted peacefully and in close contact for many years.[18] Nevertheless, although the chimpanzees are accustomed to people, they remain wary and will occasionally attack villagers if provoked.[19] The incidence of attacks has increased in recent years due to increased contact, probably as a result of rising human population pressure in the area. The challenge of maintaining the coexistence of chimpanzees with humans at Bossou is therefore significant, emphasizing the need to preserve remaining wild areas and to encourage local people to take precautions when traveling on foot alone or when carrying foods attractive to chimpanzees.

Taï Chimpanzee Project (Taï National Park, Ivory Coast)

Christophe Boesch and Hedwige Boesch-Achermann began continuous observation of chimpanzees in the Taï National Park in 1979, after a preliminary eight-month field season in 1976.[20] Establishment of a long-term project at Taï offered an important comparison population that was geographically distant and genetically distinct from the East African populations in Tanzania. Whereas Gombe and Mahale are located at the eastern edge of the chimpanzee range and their habitats include a great deal of open-canopy woodland, Taï is a dense, closed-canopy rain forest and thus offered an important window onto how ecological variation might influence social organization and behavior. Researchers at Taï have habituated four different chimpanzee groups and studied a wide range of behaviors, with notable analyses of nut-cracking with hammers and anvils, hunting behavior, and female–female relationships.

Taï National Park was designated a UNESCO World Heritage site in 1982 and is home to perhaps 4,500 chimpanzees.[21] It is the largest protected block of undisturbed primary rain forest remaining in West Africa, covering an area of 3,300 km^2 (1,274 mi^2) in southwestern Ivory Coast. The Taï Chimpanzee Project is situated on its western edge, where the terrain is comparatively flat and rainfall averages approximately 1,800 mm/year.[22] From 1982 to 1987, the primary study community ("North" group, indicated as "N" in the endnotes) included about eighty individuals occupying an area of 25 km^2.[23] In part apparently due to poaching, the community decreased steadily to fifty individuals by 1991. Following outbreaks of Ebola virus, the group declined further to twenty-nine individuals, with only two adult males, by the end of 1994. Respiratory illnesses claimed another six North group chimpanzees in 1999, with four infants subsequently dying after the disappearance of their mothers.[24] Study of a second group ("South" group, indicated as "S" in the endnotes) began in the 1980s. This group, seventy-five minutes by foot from North group, lost nine individuals to respiratory illnesses between 2004 and 2006.[25] "Middle" ("M") group, located between North and South groups, was habituated in the mid-1990s but hasn't been tracked regularly since its population declined to five individuals in 2004.[26] Systematic study of "East" ("E") group began in 2000, with two fatalities from respiratory illness reported in 2006.[27] As of 2013, population estimates for the North, South, and East groups were twenty, thirty, and thirty-five individuals, respectively.[28]

Chimpanzees are sufficiently similar to humans physiologically that disease transmission between the two species occurs. Although significant mortality from disease had already been documented at Gombe and Mahale, the dramatic nature of the outbreaks at Taï was surprising in view of the substantial differences between the sites. The Taï study groups are located deep within a vast and seemingly pristine forest, from the beginning fewer students and assistants were involved in the research, and the animals were not provisioned and thereby brought into regular close contact with humans. Nevertheless, epidemics have struck repeatedly. There is now widespread recognition of the potential risks of conducting research on wild chimpanzees, and long-term projects have instituted strict observation protocols designed to prevent the transmission of human-borne pathogens. At Taï, for example, workers and visitors are required to get vaccinations, submit to quarantine before entering the forest, and wear surgical masks when near the chimpanzees.[29]

Kibale Chimpanzee Project (Kanyawara Community, Kibale National Park, Uganda)

The ecology and conservation status of Kanyawara chimpanzees were first studied by Gilbert Isabirye-Basuta from 1983 to 1985.[30] Richard Wrangham subsequently initiated the long-term Kibale Chimpanzee Project in 1987. In a previous study of semi-habituated chimpanzees in a nearby area of Kibale (at Ngogo; see later in the chapter), female chimpanzees were observed to be more gregarious than had been reported from Gombe and Mahale, suggesting a difference between forest and woodland chimpanzees.[31] Initial analyses of social relations among Kanyawara chimpanzees revealed a pattern of social relationships that closely matched that reported in the Tanzanian populations, pointing to the need for more fine-grained analyses of the influence of ecology on social relations.[32] Since then, researchers at Kanyawara have explored a wide range of topics, including work on vocal and non-vocal acoustic communication, nutritional biochemistry, female social relationships, male aggression, and the endocrinology of social behavior.

Kibale National Park, located near the equator in southwestern Uganda, covers an area of 766 km^2. A wildlife corridor connects it with nearby Queen Elizabeth National Park to the south. Kibale includes both lowland and mid-altitude semi-deciduous rain forest (1100–1700 m.a.s.l., annual rainfall of 1400–1800 mm) and is home to 1,200 to 1,400 chimpanzees.[33] The Kanyawara study area is located on the northwestern edge of the forest, with human settlements along its border. It consists of a combination of undisturbed, selectively logged, and intensively logged regenerating forest as well as swamp forest, swamp, and grassland.[34] The Kanyawara chimpanzee study group has been comparatively stable, comprising forty to fifty individuals between 1989 and 2008, with modest growth since. It occupies an area of approximately 40 km^2.[35]

The Kanyawara chimpanzees have never been provisioned, nor have they suffered catastrophic lethal disease outbreaks. However, they do raid crops in nearby villages and are therefore at risk of injury from farmers protecting their land. They are also regularly caught in snares set in the forest by poachers to capture game animals. Snares typically cause serious wounding, including loss of limbs, if they remain tightly cinched on appendages (Plate 2). In 1997, Kanyawara researchers initiated the Kibale Snare Removal Project, headed by former park rangers, and in the first three years of operation 2,290 snares were removed from the forest.[36] The research staff at Kanyawara also sometimes intervenes to help animals injured by humans. For example, in 2009 a crop-raiding male was anesthetized and given stitches and antibiotics after being speared in the arm.[37] Similarly, animals caught in snares are occasionally immobilized and released.

Budongo Conservation Field Station (Budongo Forest, Uganda)

Vernon and Frances Reynolds conducted an eight-month study of chimpanzees in Budongo Forest in 1962. This was the second prolonged study of wild chimpanzees in East Africa, following Goodall's early work at Gombe. In contrast to Gombe, however, Budongo is a relatively large, closed-canopy forest, and research there presaged the importance of comparative studies across different habitats. In 1990, after a hiatus of more than twenty-seven years, Vernon Reynolds returned to Budongo to establish the long-term project that continues today. Research at

Budongo includes studies of ecology, social behavior, demography, and tool use, as at other long-term sites, and the Budongo chimpanzees provide an important comparative population for the study of behavioral variation across the species.[38]

Budongo Forest is the northernmost of a series of forest blocks running north to south in western Uganda. It covers an area of 352 km², constituting the largest portion of 435 km² of continuous forest in conjunction with the Siba Forest to the southeast. The habitat is semi-deciduous rain forest (1,600 mm annual rainfall) located on gently undulating terrain with an average altitude of 1,100 m.a.s.l. The forest was managed for selective timber extraction, especially of mahogany, throughout much of the twentieth century. An estimated 500 to 1,000 chimpanzees live in Budongo.[39] The study population ("Sonso" community) included about sixty-eight individuals as of 2011, occupying a territory of approximately 15 km².[40]

As at Kanyawara to the south, the "Sonso" chimpanzee community has never been provisioned, nor has it suffered lethal disease outbreaks. However, a quarter of Sonso chimpanzees have been caught in snares and traps set by poachers to capture forest animals. Two individuals at Budongo have died as a direct consequence of snaring.[41] A snare-removal project was initiated in 2000, employing local hunters. Nearly 1,000 snares were removed in the first three months of patrolling. Snares continue to be discovered, although at lower rates than at the beginning of the project. Snaring is a significant problem at other chimpanzee study sites, notably Kanyawara,[42] but it is perhaps more severe at Budongo because the forest does not have national park status, which would bring additional resources and manpower to fight encroachment and poaching.

Ngogo Chimpanzee Project (Kibale National Park, Uganda)

The ecology and social structure of chimpanzees at Ngogo were first studied by Michael Ghiglieri for a total of twenty-two months between 1976 and 1981.[43] David Watts initiated a long-term project at Ngogo in 1993, joined by John Mitani in 1995. The Ngogo study area is located in the center of the northern half of Kibale National Park, roughly 5 km from the closest park boundary. The area was never logged commercially and consequently includes a greater proportion of old-growth forest than Kanyawara.[44] Because of its close proximity to the Kanyawara study area within Kibale National Park, some 10 km away,

Ngogo has emerged as a key site for the comparative analysis of chimpanzee social behavior, demography, and feeding ecology.

The Ngogo chimpanzees utilize a range of approximately 35 km².[45] They have never been provisioned, and because of their location in the forest interior, they do not raid crops. The size of the study group has varied between about 145 and 200 individuals since 1995, making it the largest community on record.[46] Tree counts and fruit monitoring have revealed that the foods important to chimpanzees are significantly more abundant at Ngogo than at nearby Kanyawara. As a result, population density is three times greater, life expectancy is longer, and survivorship is higher at Ngogo.[47] Because of its large size, the Ngogo study group has proved a rich source of information on variation in female and male social relationships, hunting behavior, and intergroup lethal aggression.

Human disturbance has generally been minimal at Ngogo. Nevertheless, poaching for forest animals does occur, and the Ngogo chimpanzees do occasionally get caught in snares. As at Budongo and Kanyawara, the research project commits resources to anti-poaching and educational efforts to improve the conservation prospects for the chimpanzees and other wildlife in the forest.

OTHER IMPORTANT FIELD SITES

A multitude of studies have also been conducted at field sites where research has been more intermittent or shorter in duration, or where it has been difficult to regularly observe known individuals.[48] Some of these are shown in Figure 2.2. Research at all emerging sites has expanded our understanding of the extraordinary breadth of chimpanzee behavioral diversity. For example, new types of tools and tool-use behaviors have been discovered at most newly established sites, in habitats ranging from savanna-woodland (e.g., Fongoli in Senegal, Semliki in Uganda) to high-canopy rain forest (e.g., Gashaka Gumti in Nigeria) to remnant riverine forest (Bulindi in Uganda).[49]

MAINTAINING LONG-TERM RESEARCH PROJECTS

Tremendous personal dedication is required to establish and maintain a long-term research site. A steady flow of grant money is needed to fund travel and expenses for foreign workers, build infrastructure and buy equipment, pay salaries to local employees, and cover research permit fees required by host countries. This means regularly writing, submitting, and revising grant applications and managing what

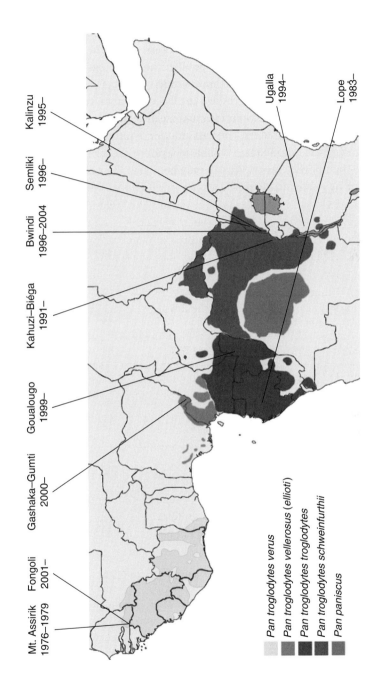

Fig. 2.2 Locations of some other important field studies. (A black-and-white version of this figure will appear in some formats. For the color version, please refer to the plate section.)

Mt. Assirik
1976–1979

Fongoli
2001–

Gashaka–Gumti
2000–

Goualougo
1999–

Kahuzi–Biéga
1991–

Bwindi
1996–2004

Semliki
1996–

Kalinzu
1995–

Ugalla
1994–

Lope
1983–

Pan troglodytes verus
Pan troglodytes vellerosus (*ellioti*)
Pan troglodytes troglodytes
Pan troglodytes schweinfurthii
Pan paniscus

amounts to a small business with responsibilities for hiring and firing employees, administering payrolls, performing quality control, keeping accounts, advising students and interns, and responding to medical requests for family members of local workers. Academic researchers who take on this challenge must publish their findings in order to continue to receive support and meanwhile meet their other institutional commitments. Beyond these logistical challenges, regular travel and spending long field seasons overseas create significant disruptions to family life for researchers. The vast amount of information we have today concerning chimpanzee social behavior and ecology is a testament to their perseverance. We can only hope that their efforts will be rewarded by a safe future for the forests in which they work.

3

Chimpanzee Fission–Fusion Social Organization and Its Conservation Implications

OVERVIEW

Chimpanzees live in mixed-sex social groups, called "unit-groups" or "communities."[1] Communities range in size from 10 to more than 200 individuals, with adult sex ratios varying between 1:1.35 and 1:4 (males:females).[2] Males live for their entire lives in the groups they are born into, whereas females generally emigrate to neighboring communities when they reach reproductive age and remain there permanently. Neighboring communities utilize discrete territories, and adult males regularly monitor the slightly overlapping boundary zones between them. Encounters between males in these areas are typically hostile. Groups of males, sometimes accompanied by females and younger individuals, occasionally make forays into neighboring territories where, if they enjoy numerical superiority, they will attack neighbors of either sex (Chapter 9). Territorial boundaries shift depending on the outcomes of these intercommunity conflicts.

In contrast to most other primates that live in stable social groups, chimpanzees do not stay in constant contact with all members of their community. Instead, individuals forage and socialize in smaller subgroups, or "parties," that usually contain fewer than ten individuals of one or both sexes and are scattered throughout their home range (Plate 3). Membership in parties, or "party composition," changes frequently, sometimes several times in the same day, and rarely remains the same for more than a few days. Most community members utilize the entirety of the community territory over the course of a year, although significant individual and sex differences exist in ranging behavior (Chapter 5). This regular reshuffling of subgroup membership and variation in ranging behavior produces a comparatively rare grouping pattern referred to as fission–fusion social organization, as mentioned in Chapter 1.

The rhythm of daily life for chimpanzees reflects a synergy between three factors in particular. First, ripe fruits, the primary food of chimpanzees, occur in comparatively small, widely dispersed patches most of the time. This means that individuals are frequently forced to forage alone or in small groups, searching widely for fruiting trees. At the same time, the need for cooperative defense of communal territory, which guarantees access to food resources, depends on the maintenance of strong social bonds among males. Males therefore spend considerable amounts of time associating and grooming together when ecological conditions permit. Finally, competition within the sexes – between females for the best feeding areas within the shared territory and between males for strategic access to females when they are ovulating – then further shapes patterns of movement and interaction within communities. Although aggregating in larger groups may benefit some primates by facilitating predator detection and avoidance, predation on chimpanzees is rare and is unlikely to affect their grouping behavior.[3]

VARIATION IN FEMALE DISPERSAL PATTERNS

In group-living primates, either males or females, and sometimes both, disperse from the group they are born into when they reach sexual maturity. Dispersal is presumably an adaptation to avoid inbreeding, reflecting an evolutionary history in which individuals that reproduced with close relatives suffered reduced fitness.[4] As already noted, male chimpanzees do not ordinarily leave their natal communities,[5] whereas females usually transfer to neighboring groups a year or two after reaching sexual maturity, between the ages of ten and fourteen years.[6] Nevertheless, not all females disperse, and occasionally some disperse more than once.[7] Individual variation in dispersal appears to reflect situation-dependent responses to the costs of leaving natal groups, underscores the potential for individual flexibility, and raises questions about mechanisms of inbreeding avoidance.

Costs of Dispersal and Potential Benefits of Non-Dispersal

The importance of nutrition for female reproductive success is well documented in chimpanzees. In captivity, where food is reliably available and nutritious, females mature faster, begin reproducing at earlier ages, and have shorter interbirth intervals than their wild counterparts.[8] In the wild, likewise, female chimpanzees maintain higher concentrations of ovarian steroid hormones, conceive more quickly, and bear the energetic burden of lactation more easily when fleshy fruit consumption is high.[9] A key challenge for dispersing females when they join a new community, therefore, is to discover and maintain access to quality food resources among unfamiliar competitors and without the help of allies. Not surprisingly, aggressive interactions between females are most common in feeding contexts, and non-feeding aggression is more common within preferred feeding areas than elsewhere.[10] Long-term data, in turn, indicate that higher-ranking females monopolize better feeding locations and have more surviving offspring than lower-ranking females.[11]

Females engage in severe aggression much less often overall than do males.[12] Nevertheless, resident females often single out immigrants as targets of threats and attacks.[13] Attacks can be sustained and violent, causing bodily harm and repulsion from the community territory. Resident females may form coalitions to attack immigrants, and they may target and even kill the infants of older females trying to enter their range.[14] For their part, females that succeed in joining a new community rank low in the social hierarchy and exhibit elevated urinary cortisol levels, an indication of increased stress.[15] New immigrants therefore tend to frequent areas far from dominant females and become more mobile in search of food during periods of fruit scarcity.[16] Together these data indicate that immigration is costly for females and prompt the question of whether individuals might sometimes fare better by not dispersing at all – a possibility suggested by the fact that females at both Gombe and Mahale that do not disperse tend to mature faster and produce their first offspring at slightly earlier ages than females who do.[17]

Several lines of evidence suggest that easier access to quality resources within a female's natal territory could favor non-dispersal. At Gombe, approximately 50 percent of females remain in the community to breed. The protected area at this site is small, sandwiched between Lake Tanganyika on one side and agricultural lands on the other, so that options for dispersal to comparably rich habitat are limited. In addition, researchers provisioned the Gombe chimpanzees with bananas for several decades. By providing a significant increase in food availability, provisioning could have directly influenced the dispersal propensities of maturing females during this time, with potential longer-term effects.[18] At Mahale, several females did not disperse during a period marked by a disease epidemic that reduced the community size by half, greatly reducing feeding competition within the territory.[19] At Kanyawara, two daughters of then-alpha females in the community remained to breed and continued to use their mothers' high-quality feeding areas.[20] And at Ngogo, where observations of non-dispersal are mounting and females appear to disperse to areas within their natal range frequented primarily by comparatively unfamiliar males, the habitat is exceptionally food rich.[21] All in all, these observations suggest that in some circumstances females can benefit nutritionally, and therefore reproductively at least in the short term, by remaining in their birth community.

The Risk of Inbreeding with Non-Dispersal

Despite the costs associated with dispersal and the potential short-term benefits of not doing so, the fact that females in most populations transfer around the time of reproductive maturity strongly suggests that the disadvantages of not dispersing outweigh the advantages. Presumably the principle disadvantage for females of staying in their natal community is that it increases the chances of mating with close relatives and incurring the fitness costs of inbreeding depression.[22] Maternally related males sometimes express mating interest toward young females once they begin to exhibit the large anogenital swellings indicative of reproductive maturity (Chapter 8).[23] In addition, since chimpanzee females typically mate with many or all males in their community, a young female is unlikely to know who her father or paternal brothers are, increasing the chances of inbreeding if they do not disperse. It may thus be difficult for non-dispersing females to avoid mating with close relatives at least some of the time.

Females that remain in their natal communities in fact appear strongly averse to mating with known relatives. At Gombe, researchers have observed females to scream loudly and flee when approached for mating by maternal brothers and mature sons.[24] Similarly at Mahale, a mother was observed to aggressively repulse her son when he showed mating interest.[25] More recent observations from Ngogo also indicate that females avoid mating with male kin.[26] Nevertheless, researchers at several sites have observed copulations between close relatives following persistent aggression by the males, confirming the potential for inbreeding if females do not disperse. In two cases of maternal sibling pairs at Gombe, the females eventually gave up resistance altogether and mated calmly and repeatedly with their brothers.[27] Similarly, a natal female at Kanyawara was observed mating with two maternal brothers.[28] Mother–son matings have been observed at both Gombe and Bossou, and presumed father–daughter pairs have occasionally been observed mating at Gombe.[29] One confirmed father–daughter pair at Gombe is known to have produced an offspring.[30]

CONSERVATION IMPLICATIONS OF VARIATION IN FEMALE DISPERSAL PATTERNS

Evidence from the seven longest-running field studies indicates that females are the dispersing sex in wild chimpanzees, and that the greatest departures from this species-typical pattern occur at the two sites most severely impacted by human encroachment. Gombe National Park protects a comparatively tiny forest area surrounded by human activity. As mentioned, roughly 50 percent of females at Gombe breed in their natal community. At Bossou, the intensity of human encroachment is even greater, such that the chimpanzee range can only support a small number of individuals, actually includes areas of human habitation, and is 6–10 km from the nearest neighboring chimpanzee community. Competition for limited resources within this small range appears to favor emigration from the community by both sexes. Surprisingly, no new females have immigrated into the Bossou group in thirty-three years of observation.[31] Thus, the long-term prospects for both of these populations are bleak.

Analyses of the immediate factors that might trigger young females to leave their natal group point to the importance of habitat quality in maintaining normal patterns of dispersal in wild chimpanzees. In the most detailed study to date, researchers at Kanyawara examined hormonal, social interaction, and ecological data collected over a fourteen-year period.[32] Young females did not exhibit hormonal indicators of increased stress before transfer. Heightened stress would be expected if these females were experiencing harassment from older residents, as occurs in some primate species when maturing females become more important competitors for resources. Although inbreeding avoidance is the likely adaptive explanation for female dispersal, it did not emerge as the immediate trigger for community transfer either. On the contrary, young females mated with community males, including relatives, before emigrating. By contrast, females appeared most likely to emigrate when fruit consumption was high, suggesting the need for high-quality nutrition before embarking on the challenging task of achieving access to a new community.

Since female non-dispersal must increase the risk of inbreeding if males do not disperse, conservation planning for wild chimpanzees needs to include identification of the minimum number of contiguous groups necessary to permit normal dispersal. This, in turn, will then make it feasible to estimate the amount of undisturbed habitat needed to ensure the genetic viability of regional populations.[33] In the absence of sufficient territory, the long-term prospects for communities are uncertain. Even at Gombe, where a long-term research presence and intensive conservation efforts have offered important protection for the chimpanzees there, the ultimate health of the park population is still jeopardized by human encroachment. Similarly, the long-term survival of the Bossou community now depends entirely on the tolerance and goodwill of the local people with whom they must share their territory. Continued research identifying the key connections between community structure, habitat quality, and range size will be essential if chimpanzee conservation efforts are to succeed.

4

Sex Differences in Ranging and Association Patterns

OVERVIEW

The fission–fusion social organization that characterizes all chimpanzee communities is immediately recognizable when observers attempt to collect data on known individuals over extended periods of time. On a typical day, a researcher might begin by following a mother and her dependent offspring, accompanied by several adult males, as they make their way to a fruiting tree (Plate 4). At some point, after feeding in and then leaving this tree in search of another feeding location, one of the males might wander off, for no obvious reason or perhaps in the direction of vocalizations heard from another part of the forest, not to be seen again for a week. Alternatively, the five individuals might be joined by another small group like theirs and continue together for the rest of the day, or even several days. Or the initial small group might join a noisy group of twenty individuals feeding in a grove of trees laden with ripe fruits and remain with them, first feeding and then resting and socializing. Toward the end of the day, the mother and her offspring might leave the group to find their own spot to sleep for the night. And so on. The permutations are infinite and generally unpredictable.

A combination of ecological and social factors is implicated in this fluid, ever-shifting grouping pattern. Ecologically, the availability of ripe fruits sets the baseline conditions for ranging and association. Individual trees that are important sources of ripe fruit for chimpanzees can flower in different months in different years and may or may not bear fruit synchronously with other trees of the same species (Chapter 1, endnote 21). Edible fruits therefore occur in discrete and unpredictable "patches" of varying sizes scattered widely throughout community territories. A patch might consist of a single, isolated tree in fruit or a cluster of trees fruiting at the same time. Consequently, unlike temperate habitats where all members of a tree species may flower and fruit nearly simultaneously with reliable seasonality, in chimpanzee habitats the timing and location of fruit production is extremely variable. Chimpanzees are therefore forced to search almost constantly for new feeding areas, often monitoring known trees as they go.[1] The size and distribution of food patches then influence the size and membership of foraging subgroups.[2]

Significant sex differences in gregariousness combined with individual preferences, apparently, for affiliating with or avoiding specific group mates are then overlaid on the baseline circumstances of ripe fruit availability. In general, females travel and socialize with other adults less than males do. An individual female's mobility and sociability are influenced by her age, status, reproductive condition, whether she has dependent offspring, and the demographic structure of the community (see later). Male chimpanzees, by contrast, frequently associate with adults of both sexes. Males aggregate in larger parties in the presence of swollen females (Chapter 7), when engaged in territorial defense (Chapter 8), and when hunting monkeys (Chapter 9). Together with sex-specific patterns of aggression and affiliation (Chapters 5 and 6), these differences in territory use and gregariousness point to fundamental asymmetries between the social lives of male and female chimpanzees.

FACTORS INFLUENCING AN INDIVIDUAL'S RANGING AND ASSOCIATION BEHAVIOR

Sex Differences

Adult chimpanzees range anywhere from a few hundred meters to many kilometers daily, averaging between 1 and 5 km per day across sites.[3] At all sites, males travel more widely than females, utilize the majority or all of the community range over the course of a year,[4] and patrol boundary areas regularly (Chapter 8).[5] Females with offspring move more slowly than adult males,[6] are likely to suffer greater travel costs in terms of reduced feeding time,[7] and may be more vulnerable than males to intercommunity aggression when visiting the periphery of the community range. Thus, females generally concentrate their foraging effort in more restricted but overlapping core areas that are stable over long periods of time.[8] Nevertheless, both male and female ranging patterns can be quite variable. Males may occasionally restrict their ranging to especially familiar areas to maximize foraging efficiency when food is generally scarce.[9] Conversely, female core areas are not strictly defended territories, and females may wander considerably depending on the availability of ripe fruits throughout the community range.

Overall, adult females tend to be less sociable than adult males,[10] although variation exists in average female gregariousness both within communities over time and between sites. At Gombe from 1972 to 1975, for example, females spent about 65 percent of their time alone (which includes when they are accompanied by an infant), whereas males were alone only 27 percent of the time.[11] Averaged over the period from 1975 to 1992, however, Gombe females and males spent about 24 and 10 percent of their time alone, respectively,[12] and from 1995 to 2004, time alone for individual females ranged between 39 and 48 percent annually.[13] Thus, while the scattered distribution of food resources forms the basis of chimpanzee fission–fusion behavior generally, it is also likely that habitat fluctuations influence individual levels of gregariousness in the shorter term. This might explain why at Taï, where the habitat is richer and less variable, time spent alone by females has remained fairly constant over the years at about 18 percent.[14]

Sources of Variation among Females

Despite ranging less widely and spending more time foraging alone than males on average, many females nonetheless associate with some community members regularly and utilize most areas of the community range from time to time. Grouping patterns in chimpanzees are typically characterized in terms of an "index of association," which is calculated as the number of times in which two individuals are observed together divided by the total number of times each is observed overall.[15] Observations from all long-term sites indicate that, on average, male–male pairs associate the most, and male–female pairs associate somewhat more than female–female pairs.[16] Nevertheless, dyadic association values for particular female–female pairs in some study groups are as high as or higher than the highest male–male values.[17] Thus, low average gregariousness does not imply that females are inherently less capable than males of being social. Instead, it results in part from the presence of a higher proportion of relatively asocial females compared with males in a community.

While the quality and distribution of preferred foods set the baseline conditions for general patterns of female ranging and association at different field sites, two additional factors in particular affect individual female variation within communities: reproductive condition and social status. Unusual demographic characteristics can also have a profound effect on females. These include severe population reduction as well as sub-structuring within extremely large communities, both of which are discussed in the following sections.

Reproductive Condition

A key influence on a female chimpanzee's ranging and association behavior is her reproductive state. This is manifested in two ways. First, prior to dispersing to a neighboring community, adolescent females who have begun menstrual cycling spend less time with their mothers, brothers, and previous close male associates and more time with other adolescent and adult males.[18] At this time, adolescent females also often visit one or more neighboring communities temporarily while they are in estrus,[19] and consequently they may spend more time alone than formerly as they wander between groups.[20] As discussed in Chapter 3, although aggression between females is generally rare and mild, resident females threaten and attack visiting and immigrating females at comparatively high rates. Presumably as a consequence, new immigrants initially associate preferentially with resident males, who frequently intervene to prevent resident females from harassing them.[21]

Once she is established in a new community, a female's ranging and association behavior continues to be strongly influenced by her reproductive state. The average menstrual cycle length of female chimpanzees is about thirty-six days, and the interval around ovulation is marked by a hormonally stimulated, gradual, and ultimately dramatic swelling of the anogenital area (Chapter 7). After an initial swelling period of roughly seven days, swellings peak in size twelve to thirteen days before ovulation, and then quickly decrease over several days following ovulation.[22] Thus, swellings provide an important indicator of female ovarian function, are highly attractive to males, and are easily monitored by researchers.[23] In those studies where cycling and non-cycling females have been analyzed separately, swollen females travel more widely and associate more regularly with males compared with anestrous females.[24] Additionally, swollen females sometimes range alone with a single male on mating "consortships" for several days at a time before rejoining other group members (Chapter 7).[25] Once they become pregnant, females then reduce their ranging compared even to females with nursing offspring, perhaps allowing them to conserve and store energy for the subsequent demands of lactation.[26]

Social Status

As discussed in Chapter 3, females compete for access to food resources and establish overlapping core areas where they forage preferentially. Spatial clusters of core areas, in turn, are referred to as "neighborhoods," and different neighborhoods can exhibit broad differences in habitat quality. An important consequence of neighborhood differences in fruit availability for patterns of ranging and association is that high-ranking females who compete successfully for core areas in more fruit-rich areas interact more frequently with nearby females, as well as with males, than do low-ranking females.[27] In addition, higher-ranking females can more easily tolerate the costs of gregariousness when obliged to forage together during periods of relative fruit scarcity.[28] By contrast, lower-ranking females, which includes new immigrants, tend to settle in peripheral areas of the community range where they may need to travel further in search of more sparsely distributed ripe fruits.[29] Consequently these females associate less frequently with community members of either sex, with the exception of nulliparous females (females yet to reproduce), who can be as gregarious as adult males.[30]

EFFECTS OF EXTREME DEMOGRAPHIC CONDITIONS

Greater Cohesion in Smaller Communities

In demographically unusual communities, ranging and association behaviors of both males and females can differ substantially from more typical patterns. When group size and the number of males are greatly reduced, communities appear to become more cohesive. This has been observed in two West African communities especially impacted by humans. At Bossou, the community is both very small and isolated from other groups, situated within an agricultural area densely populated by humans. Between 1993 and 1999, all members of the community were observed together in a single party on 26 percent of observation days, and at least 75 percent of the community formed a single party on 36 percent of observation days.[31] Similar results were reported in earlier periods of the long-term study.[32] At Taï North, likewise, after the majority of community males died during disease epidemics, females associated with the remaining males and with each other more than previously.[33]

It is unclear why these smaller communities with fewer males are more cohesive. Generally relaxed feeding competition could reduce the costs of associating together, especially where crop raiding offers dense nutritional resources.[34] However, since larger feeding parties would still deplete particular food patches more quickly, there would continue to be pressure on individuals to forage alone or in small groups.[35] Alternatively, increased group cohesiveness could be useful for defense against predators or aggressive neighbors. Unfortunately, there are no data available to test these possibilities at Bossou or Taï. Finally, adult males in all populations direct substantial amounts of aggression toward females and their offspring (Chapter 6). Increased cohesiveness in these smaller communities could simply result from having fewer potentially annoying males around. In support of this idea, a study at Kanyawara showed that mothers were less likely than non-mothers to associate with males, although mothers and non-mothers were equally likely to associate with other females. In addition, mothers stayed closer to their infants when in the presence of males and let their infants move farthest from them when there were relatively more females in social groups.[36]

Sub-Structuring in Large Communities

With roughly 200 individuals but occupying a territory no larger than nearby Kanyawara, the Ngogo community is the largest chimpanzee group on record. An early study showed that the approximately thirty-five adult and adolescent Ngogo males could be divided into two subgroups based on a statistical analysis of membership in temporary parties. The bigger subgroup, with twenty-three individuals, comprised higher-ranking and older individuals. While there was extensive overlap in range use between the two subgroups, the larger one visited all areas of the territory, whereas the smaller one utilized only about 90 percent of the community range. This pattern of ranging and association was stable over many years but does not yet appear to reflect a schism within the community that could lead to a permanent fission event, as occurred at Gombe in the 1970s (Chapter 8). A possible explanation for this sub-structuring is that it may be difficult for young males to join the unusually large adult male network at Ngogo, forcing them to remain somewhat separate until they can gradually integrate themselves into the adult hierarchy.[37]

An additional effect of large group size at Ngogo is that although males use the entire community territory, individuals nonetheless tend to spend more time in some areas compared to others. The Ngogo habitat is exceptionally rich in preferred foods for chimpanzees, perhaps making it feasible for males to range somewhat selectively.[38] As a consequence, males associate disproportionately with females that use similar parts of the territory, a tendency that can then persist when they forage in other areas of the range as well. Although higher-ranking males at Ngogo sire more offspring than lower-ranking males, as is the case at other sites (Chapter 7), male–female pairs that display these socio-spatial relationships often reproduce together, to the extent that range use can have as significant an effect as rank on male reproductive success. In the most extreme case, one male sired three of the same female's four offspring.[39] Such male–female association patterns, nested within a larger multi-male/multi-female group, have not been observed in other communities and point to the potential for habitat characteristics to influence the social structure of a community.

On the Threshold of a Community Formation Event?

Variation in ranging and association patterns within and between field sites provides excellent opportunities for investigating the relationship between chimpanzee ecology and grouping behavior. As discussed in Chapter 3, conservation efforts will depend on the success of such work. At the same time, socioecological research is also fueled by the suspicion that rare social developments, which can have far-reaching effects on chimpanzee community life, await discovery. For example, community extinctions at Gombe and Mahale were once-in-a-lifetime occurrences that had profound consequences for the individuals in those populations (Chapter 8). The unusually large Ngogo community may experience a similarly pivotal upheaval, as it appears to have the potential to fission permanently. While group fission would raise the possibility of violent conflict between individuals that were once members of the same community, given the size of the Ngogo group, it could alternatively result in the creation of a new and permanent community, an event never before documented in the wild.

5

Female Social Relationships

OVERVIEW

A striking feature of wild chimpanzee social existence is how different the lives of adult females and males are. As in other primate species, relationships within and between the sexes are greatly influenced by sex-specific dispersal patterns. As the transfer sex, female chimpanzees upon maturity must forge completely new relationships with members of a neighboring community. Throughout their adult lives, female social behavior is then dominated by mothering, competition for resources, and often tense interactions with adult males. Males, by contrast, spend their entire adolescent and adult lives enmeshed in a complicated web of relationships with each other, relationships that are at once cooperative, competitive, and ever-shifting. As a consequence of these divergent social trajectories, the frequencies at which friendly and unfriendly behaviors are given and received among and between females and males vary tremendously.

Sex differences in behavior appear in infancy.[1] For example, both female and male infants at Kanyawara use sticks in a number of different contexts: during play, during aggressive interactions, when probing holes for water or honey, or when otherwise occupied with their routine behaviors. They may carry them for hours at a time in their hand, under their arm, or wedged between thigh and abdomen. But whereas females more than males simply carry sticks, sometimes taking them into day nests and resting with them in a way reminiscent of mothering behavior, males more than females use sticks in aggressive contexts.[2] In the social sphere, male infants at Gombe begin to travel independently earlier, maintain further distances from their mothers starting at age three, and spend more time in social play than female infants. Male infants also interact more with adult males than female infants do, even though there is no difference between infant females and males in how often they inter-act with adult females.[3] These early sex differences foreshadow important differences between adult females, who concentrate on offspring care and resource acquisition, and males, who devote more energy to social interactions important for navigating the male status hierarchy.[4]

Young male and female chimpanzees can also differ in their predispositions to learn particular adult activities, laying the ground-work for later distinctions in adult behavior. For example, infant females at Gombe spend more time than infant males watching their mothers "fish" for termites with small stick tools (Chapter 11), whereas males spend more time playing around the termite mounds while their mothers fish.[5] Infant females then use techniques (e.g., tool length and dip depth) more similar to their mothers' than do males, succeed in acquiring termites with a fishing tool more than two years earlier than males on average, and gather more termites per "dip" than males who have been termite fishing for equally as long. Juvenile and adolescent females at Gombe continue to dip for termites more frequently than males. At Mahale, likewise, young females dip for ants with tools more than males do.[6]

Early sex differences in learning these tool-assisted foraging techniques foreshadow significant sex differences in adult foraging behavior and nutrition. Adult females at both Gombe and Mahale rely more on insects as a source of animal protein, whereas males focus on mammalian prey and often hunt in groups with other males (Chapter 8).[7] Moreover, divergent learning patterns in female and male infants appear to depend on the presence of sex differences in adult behavior. At Bossou, for example, where there are few males in the community and hunting is rare, adult females and males do not differ in the amount of time spent dipping for ants with tools. Correspondingly, infant females and males at Bossou do not differ in the amount of time spent observing their mothers ant-dip or in the age at which independent ant-dipping behavior appears.[8]

As young chimpanzees mature, sex differences become more pronounced, and their behaviors start to resemble those of same-sex adults. Males begin to range separately from their mothers in early adolescence, sometimes alone but increasingly in the company of adult males and swollen females. As they mature, they receive increasing amounts of aggression from older males until they become integrated into the adult male hierarchy.[9] Late adolescent males begin to direct threats and attacks toward adolescent and adult females, ultimately establishing dominance over each by adulthood.[10] Immature females, by contrast, are less gregarious than their male age-mates and remain close to their mothers throughout adolescence. Late adolescent females typically begin ranging alone and associating with males when they start to have small estrous swellings and approach dispersal. By the time they reach adulthood, females on average exhibit sharply different patterns of social behavior from male chimpanzees. In particular, sex differences in the frequency of friendly and aggressive behaviors directed toward and received from group members reveal fundamental sex differences in social bonding. Typical tendencies in adult females are described in the following sections. Male behavior patterns are addressed in Chapter 6.

FRIENDLY BEHAVIORS

Toward Offspring

The primary affiliative relationships of adult females are with their dependent young (Plate 5). The years-long, close physical relationship between mother and offspring invites this conclusion, and patterns of grooming confirm it. In all long-term study populations, females groom most with their offspring, next with adult males, and least, if at all, with other females.[11] The close bond between mother and daughter lasts until dispersal, after which the two become de facto strangers.[12] When females do not disperse from their natal community, they sometimes maintain close relationships with their mothers throughout adulthood. At Gombe, though not at Mahale, parties containing mothers, daughters, and the dependent offspring of both are common, and mothers and adult daughters continue to groom regularly.[13] Similarly, at several sites there is evidence that mothers and adult sons associate and groom with each other preferentially, even though sons are dominant over their mothers and sometimes are aggressive toward them.[14]

Toward Males

Although adult females often travel, feed, and rest in proximity to unrelated adult males, they interact comparatively rarely with them in obviously friendly ways. Most significantly, as mentioned previously, adult females generally groom with males much less frequently than they groom with their offspring. More commonly, female behaviors toward nearby males fall on a spectrum between overt submissiveness and peaceful coexistence. Nevertheless, substantial individual variation exists. In one study at Gombe, for example, four females with offspring devoted between 76 and 96 percent of their grooming time to their offspring, between 4 and 18 percent to males, and between 0.1 and 16 percent to other females. A fifth female, however, devoted 49 percent to males, 47 percent to offspring, and 3 percent to females. A sixth female with no offspring devoted 60 percent of her grooming to males and 40 percent to females.[15]

Variability in grooming patterns underscores the individualized nature of social relationships among chimpanzees, as well as the diverse functions served by grooming – as a nurturing behavior, a means of relieving stress, or a mechanism for establishing short- or long-term bonds. For females, grooming with adult males offers several potential benefits. First, it can constitute a component of greeting behavior between individuals reuniting after a period of separation.[16] Grooming in this context appears to reduce the likelihood of immediate attack by males, who are frequently aggressive during reunions (Chapter 6, endnote 28). Second, grooming frequently occurs at least briefly in the context of sexual behavior (see later). Finally, males may protect females that direct friendly behaviors toward them, occasionally intervening in their behalf when they are harassed by other females. This is especially important when females emigrate to a new community, where they frequently receive aggression from resident mothers (Chapter 4).[17]

A female's reproductive state significantly influences her grooming patterns with males. Observations at most sites have shown that swollen females in group contexts groom with males more than usual, in part a reflection of male mating effort since bouts are more often initiated by males.[18] Similarly, grooming with males is more frequent than usual during "consortships," when a male and a swollen female wander together for several days at a time apart from other group members (Chapter 7). Consortships occur when a male succeeds in leading an often unwilling female away from social groups where she would typically mate with many, if not all, of the males present. Persuading females to accompany them sometimes requires coercive aggression by males.[19] A percentage of matings at all study sites takes place during consortships, and grooming may have the immediate effect of calming anxious females as well as potentially aggressive males.

Toward Other Females

Where females do associate together, their relationships usually appear better characterized as mutually tolerant rather than bonded. For example, "nursery parties" of two or more adult females with dependent offspring have been observed at all sites since the earliest field studies.[20] In contrast to males that associate together, however, the adult females in these groups often do not appear particularly interested in each other, mainly resting alone or with their young after bouts of feeding rather than engaging as males do in long grooming sessions.[21] Thus, rather than associating for each other's company per se, it seems more likely that mothers are attracted to other females in these situations for the benefit of their offspring. In support of this idea, studies at Gombe have found that juveniles tend to play most with youngsters close to them in age,[22] and that mothers with juvenile, but not infant, offspring are most likely to associate with other females who also have dependent juveniles.[23]

Nevertheless, although females on average groom together less than males, many do establish enduring relationships with each other, as revealed by comparatively frequent grooming interactions and other supportive behaviors. Reports from all field sites except Kanyawara indicate that at least a few unrelated females at each have been observed to groom with one another to a significant extent, albeit less so than males do.[24] In addition, unrelated adult females sometimes form temporary coalitions in aggressive contexts. Females at five of the seven long-term sites have been reported to join together to attack other females, sometimes causing serious injury to victims and their offspring.[25] At several sites, females have been observed to come to the aid of other females under attack from males or other females.[26] Females sometimes embrace or engage in other forms of friendly contact with other females in stressful contexts such as reunions or after aggressive incidents.[27] Finally, females occasionally share food with each other.[28]

AGGRESSION, DOMINANCE, AND SUBMISSION

Adult female chimpanzees direct aggression toward other group members much less frequently on average than males do.[29] Their aggressive behaviors are also generally much lower in intensity than those of males, consisting primarily of threats and supplants. Females are occasionally observed directing severe aggression toward other females and their offspring, with probable long-term consequences for access to critical feeding areas (Plate 6).

Toward Offspring and Unrelated Youngsters

The close relationship between mother and infant is immediately obvious in the field, revealed by high levels of physical contact, grooming, food sharing, and, by the mother, protection. Nevertheless, discord that can lead to aggressive behavior may develop as offspring grow older. Two types of conflict in particular have been reported. First, some infants vigorously resist their mother's efforts to prevent nipple contact during weaning, prompting mothers to react aggressively in turn. Potential aggressive behaviors by the mother range from threatening vocalizations to hitting, biting, and kicking.[30] Second, mothers sometimes share food with their young, particularly hard-to-process plant items that they permit infants to take from their mouths. As infants age, however, mothers share less. At Gombe and Mahale, mothers have been reported to threaten, supplant, and even take food from their offspring at food sources.[31] Meanwhile, they are extremely protective of their dependents and sometimes threaten unrelated youngsters who interact roughly with them.[32]

Toward Males

Adult females at all study sites are subordinate to all adult males and rarely direct aggression toward them on their own, though they may retaliate against late adolescent males that are in the process of establishing dominance over them.[33] When females do attack adult males, it is typically in order to protect an infant or to come to someone else's aid.[34] Taï is the only site where females seem to get involved comparatively often in fights between adult males. In a two-month period, one female twice came to the aid of her long-time male ally who was under attack; in a third case a different female formed a temporary coalition with a male to come to the aid of a second male under attack. During this same period, four instances were recorded of a female joining a male to attack another male, temporary coalitions that involved three different females and two different males.[35] Several cases of females joining in male attacks on other males have also been reported from Mahale.[36]

Toward Other Females

Aggressive interactions between adult females in the same community are comparatively rare.[37] As a consequence, it is often impossible to determine dominance relationships between them. In two studies at Mahale and one at Taï, linear dominance hierarchies were identified for adult females.[38] By contrast, females at Gombe in three time periods and at Kanyawara in the early years of the project could only be assigned to high, middle, and low rank classes, with a likely alpha female.[39] Aggressive interactions between females were too rare at Kanyawara a decade later, and also too rare in studies at Ngogo and Budongo, to draw conclusions.[40] In recent years, however, incidents of female–female aggression at Kanyawara have been recorded more often, making it possible to assign female dominance ranks.[41]

Despite its relative infrequency, female–female aggression is likely to have significant adaptive consequences. Aggressive interactions between females are most frequent in contexts of short- and long-term competition for plant food resources, and studies at Gombe and Kanyawara have shown that the core areas of higher-ranking females include higher densities of preferred fruits. A companion study at Kanyawara also indicated that better diets were associated with improved ovarian function (Chapter 3, endnotes 8–10). Further analyses in these two communities have shown that lower-ranking females that are lactating receive more aggression from males and have higher levels of circulating steroid hormones indicative of stress.[42] Received aggression can interfere with feeding effort, and stress can negatively affect reproductive function. Overall, higher-ranking females at Gombe and Kanyawara have more surviving offspring and the spacing between their births is shorter, and at Gombe the daughters of high-ranking females mature at earlier ages, all of which would contribute to greater long-term reproductive success.[43]

Extremely violent aggression between females is rare but nevertheless has been observed in most communities. Attacks often target infants as well as mothers, and infant deaths as a consequence of such incidents have now been documented at four study sites (Chapter 8). Severe attacks can be divided into four categories, and infanticides have been observed in each: (1) attacks by coalitions of two or more females on strangers or recent immigrants,[44] (2) attacks by coalitions of two or more females on resident females and their offspring,[45] (3) one or more females joining in with males attacking stranger or new immigrant females,[46] and (4) one or more females joining in with males attacking resident females and their offspring.[47] Since infants cling ventrally to their mothers during attacks and mothers vigorously defend them, aggressors sometimes coordinate their efforts to seize offspring, with one or more individuals restraining the mother while another grabs the young victim.[48] The frequency of such attacks on immigrants, who will increase feeding competition within a finite community range, emphasizes the importance of female–female aggression in determining access to resources.

THE PROFOUND INFLUENCE OF REARING YOUNG

The period of infant dependency in chimpanzees is long. Although researchers have documented variation in female social relationships both within and between communities, the daily life of a female is nonetheless shaped profoundly by nurturing and protecting her off-spring. Long, usually peaceful days traveling, holding, grooming, and playing with offspring are the rule for most females. Competition between adult females for access to resources explains the occurrence of aggressive interactions between them, but these incidents are infrequent and generally mild. Episodes of severe aggression between females are even rarer, no doubt because the immediate benefits are comparatively low and the risks, especially if carrying a vulnerable clinging infant, are great.[49] Meanwhile, females often avoid associating with males, who can exhibit explosive and seemingly unpredictable bursts of aggression, of which females and their offspring may be either indirect or direct victims. This is considered in more detail in the following three chapters.

6

Male Social Relationships

OVERVIEW

The social lives of chimpanzees are influenced by how and with whom they compete for the resources necessary to their survival and reproduction. For females, this primarily means vying with other females for feeding locations within the community territory. In general, female–female competition occurs on a one-to-one basis, with individuals prevailing over or submitting to each other based on infrequent and low-level aggressive interactions. Although females are at times overtly friendly with one another and occasionally may form temporary coalitions to attack other females, on a daily basis they are largely indifferent to each other. Instead, adult females devote most of their social energy to nurturing and protecting their dependent offspring, who will forage, rest, and sleep with them for many years. Thus, female social life is characterized mainly by intensely supportive relationships with close relatives, competitive though occasionally friendly associations with other females, and deferential relationships with adult males.

Male existence, by contrast, revolves around both within- and between-community rivalry with other males for status, territory, and access to sexually receptive females. More so than for females, consequently, their social lives are shaped by a complicated mixture of competition and cooperation. On the one hand, males regularly patrol and defend territorial boundary areas together. Violent and sometimes lethal territorial aggression, in which several males from one community cooperate to attack isolated individuals from a neighboring community, has been observed at all study sites (Chapter 8). Meanwhile, the same males that join forces in dangerous territorial interactions also compete vigorously against each other for dominance status within their own communities. Although dominance conflicts are usually resolved through one-on-one aggressive interactions, they too can involve coalitions, are sometimes extremely violent, and can be lethal. Thus, relationships among males embody a delicate balancing act, with individuals committing large amounts of time to friendly interactions while also intermittently engaging in escalated aggression with each other to achieve dominance status.

Male behaviors toward females, on the other hand, are oriented more toward gaining long-term sexual access than creating supportive social bonds. Males form coalitions much less frequently with females than they do with males and, correspondingly, spend less time directing affiliative behaviors toward females. Meanwhile, despite the fact that their relative ranks are never in question, males frequently direct seemingly unprovoked aggression toward both cycling and non-cycling females. This conspicuous feature of chimpanzee social life at most field sites has been interpreted as a form of indirect sexual coercion (Chapter 7) and is indicative of the profound difference that exists between male–male and male–female relationships. Females, for their part, often appear wary of males, monitoring their movements constantly and often avoiding close proximity with them. Nevertheless, males and females do engage regularly, if comparatively infrequently, in calm and friendly contact with each other, and some individuals, especially mothers and sons, maintain friendly relations over extended periods of time.

FRIENDLY BEHAVIORS

Toward Infants, Juveniles, and Adolescents

Adult males interact comparatively infrequently with infants and juveniles. Since sexually receptive females typically mate with most males in their community, males are unlikely to know who their offspring are and provide no parental care. Nevertheless, although there are few systematic studies of males and young, anecdotal evidence indicates that males are often generally friendly toward them when associating in mixed groups. For example, even though adult males groom with infants and juveniles much less than with adults,[1] they occasionally wrestle gently with them or tickle them.[2] In addition, they are tolerant of youngsters in situations that would provoke aggressive behavior if older individuals were involved. Young individuals preoccupied in energetic play can bump into adult males, or even intentionally rumble over them, without response. Similarly, infants and juveniles that interfere in sexual interactions between their mothers and adult males, climbing on either or both during copulations, are typically ignored. In rare cases, adult males have even been known to "adopt" orphaned infants, traveling with them, providing protection, and helping them acquire food.[3]

Toward Other Males

In addition to being generally more gregarious than females (Chapter 5), adult male chimpanzees routinely engage in affiliative behaviors with each other. Reports from all long-term study sites indicate that on average males groom with each other much more often and for longer periods of time than they groom with females, although some males may show a preference for grooming with females.[4] Grooming can constitute more than 25 percent of a male's waking time, and a single bout can last more than an hour.[5] In contexts of social excitement with other group members or aggressive interactions with neighbors, males frequently touch, embrace, and mount each other (Chapter 10).[6] Such contact behaviors appear comforting, an interpretation supported by data from captivity showing that friendly physical contact can reduce behavioral indicators of anxiety (e.g., self-scratching) as well as circulating levels of hormones associated with stress (Chapter 1).[7] Males at all sites hunt, and successful hunters often share meat from their kills with other males (Chapter 9). Finally, adult males may even wrestle together playfully on occasion.[8]

Pairs of males that regularly exchange affiliative behaviors develop enduring, mutually supportive relationships. Few studies have tracked the stability of specific dyadic relationships over long periods of time, but initial analyses indicate that pairs of males can maintain bonds for at least ten years.[9] Grooming is the most commonly observed, and evidently the most important, friendly behavior between such close pairs, with partners regularly alternating roles as groomer and groomee.[10] Researchers at Gombe, Mahale, and Taï have inferred that males receive coalition support in return for grooming as well as sharing meat,[11] and analyses at Taï and Ngogo have supported this conclusion quantitatively.[12] Since temporary coalitions can play a decisive role in aggressive interactions related to status competition, sexual access, and conflicts with neighboring groups, establishing and maintaining close social bonds through affiliative behavior thus plays a key role in the lives of males.[13]

The exchange of friendly behaviors does not occur uniformly across all males, reflecting the complicated nature of male social relationships. This is particularly evident for grooming. In general, males tend to groom "up the hierarchy." That is, younger and lower-ranking individuals tend to initiate grooming bouts more than higher-ranking ones, and they groom their superiors more than the reverse.[14] The implication of this upward directionality in grooming behavior is that high-ranking males are attractive as potential allies whose support is worth courting. However, the repeated observation that alliances between individual males can shift, and that males can act in ways that disrupt alliances between others (e.g., see "separating interventions" later), implies the potential for a significant degree of strategic social maneuvering among males. Thus, alpha males at some sites initiate more grooming than their subordinates once they have climbed to the top of the social ladder, reversing the typical directionality of grooming.[15] This appears to represent an effort to maintain the support of subordinates while inhibiting the formation of alliances between potential challengers.[16]

It is unclear why particular males ultimately develop close relationships with one another. We might expect that the strongest bonds would form between close relatives, since relatives can increase their inclusive fitness via kin selection.[17] The best candidates would be maternal brothers near in age (paternal half-brothers might well be extremely close in age, but males are unlikely to be able to recognize them; Chapter 7), since the prime competitive years for a male may last only until his early thirties. Thus a fifteen-year-old entering the adult competitive arena would benefit most from a brother within fifteen years or so of his age. However, the long interbirth intervals typical of chimpanzees, combined with intervening female births as well as deaths from disease or aggression, can easily result in wider age differences between full or maternal half-brothers.[18] Moreover, older males will already have established relationships with other males and may not be available for younger siblings. Thus, strong bonds between brothers appear to occur only rarely.[19] In the absence of similarly aged and recognizable kin, males then establish their strongest affiliative relationships with unrelated individuals,[20] often of comparable age and rank.[21]

Toward Females

As previously discussed, males generally groom less with females than with other males, although they may increase their grooming time with females that have estrous swellings (Chapter 5, endnote 18). At Taï and Kanyawara, for example, males on average devote only about a quarter of their grooming time to adult females, and at Ngogo and Budongo they devote roughly 10 and 15 percent, respectively.[22] As mentioned in Chapter 5 (endnote 14), adult sons sometimes groom to a considerable extent with their mothers. With regard to other affiliative behaviors, males frequently share meat with females (Chapter 9) and occasionally play with them.[23] Some males have been observed to form coalitions comparatively frequently with particular females, although this does not appear to be commonplace.[24] Likewise, males occasionally intervene in support of females during aggressive interactions, particularly on behalf of new immigrants being harassed by resident females (Chapter 4, endnote 21).

AGGRESSION, DOMINANCE, AND SUBMISSION

Males at all study sites are physically aggressive much more frequently than females, and they target females and males alike. For example, in a one-year study at Kanyawara, males on average were observed to attack, chase, or display at other individuals roughly once every three and a half hours, fourteen times as often as females.[25] Similar figures have been reported from Gombe.[26] Higher-ranking males tend to be responsible for more aggressive acts than their subordinates, while subordinates refrain from displays of aggression in the presence of higher-ranking individuals unless their dominance relationships are unstable.[27] Meanwhile, late adolescent and young adult males typically harass females before attempting to challenge older males.

Aggression by males is especially common in three contexts. First, aggression often occurs when individuals meet up after intervals of separation, episodes referred to as "reunions." Reunions are a regular feature of fission–fusion social organization and are often characterized by extreme social excitement, with adult males performing generalized charging displays or chasing particular individuals before everyone settles down to a period of relaxed feeding, grooming, or resting. A likely explanation for the frequency of aggression at reunions is that males are motivated to reaffirm their dominance status after periods during which they are unable to monitor each other's behavior.[28] Second, aggression is frequent when males are in the presence of swollen females (Chapter 7). Finally, aggression is also common when individuals vie for shares of especially valuable foods, such as meat after successful hunts[29] or honeycomb after a bee's nest has been raided.[30]

Toward Infants, Juveniles, and Adolescents

Adult males direct comparatively little severe aggression toward infant and juvenile community members. Instead, youngsters may occasionally receive mild threat gestures from adult males if playing or feeding too close. They may also be incidental victims of attacks on their mothers or get caught in the fray of generalized charging displays. As young individuals progress through adolescence, adult males become increasingly aggressive toward them, especially late adolescent males that begin to seek out their company.[31] Nevertheless, although adult males are generally tolerant of youngsters, lethal attacks on infants have been observed at all long-term study sites (Chapter 8). While such events are rare, it is likely that mothers are less gregarious than non-mothers, and stay close to their infants when they are around adult males, in part because adult males are potentially dangerous.[32]

Toward Other Males

Adult males are aggressive toward other males of all ages, although they direct the most acts toward other adults.[33] Physical conflicts can involve grappling, biting, hitting with the hands, and stomping on prostrate victims with the feet.[34] When more than two males gang up on an opponent, some may hold the victim down while the others beat and bite him. Altercations are usually brief, but in the most violent cases may last for twenty minutes or more.[35] Severe fights can result in serious injuries and even death (Chapter 8). Indeed, in addition to recognizing individuals by their characteristic facial features, body proportions, coloring, and general demeanor, researchers often identify their study animals by deformities resulting from past conflicts: facial scars, missing fingers or toes, or partial ears.

Males threaten and fight with each other to compete for status. The outcome of repeated aggressive interactions is the establishment of dominance relationships in which one individual, the subordinate, is likely henceforth to defer to the other in competitive contexts. Once a dominance relationship is formed, the two individuals engage less often in physical aggression. Subordinates then produce the chimpanzee's characteristic vocalization of submission, the "pant grunt," toward dominants at reunions or other moments of social excitement. Combining records of aggressive acts and pant grunting, researchers at most field sites have revealed linear dominance hierarchies among adult males.[36] In other communities, aggressive interactions are sufficiently rare that researchers can only group males into high, middle, and low rank categories, and cannot distinguish ranks within these groups beyond the presence of a clear alpha male among the high-ranking males.[37]

A common feature of male aggressive behavior is the formation of temporary coalitions and long-term alliances.[38] Active support from coalition partners can help a dominant individual maintain his status or, equally, tip the balance in favor of a challenger.[39] Consequently, males apparently attempt to prevent the formation of close relationships between challengers and their potential allies by physically disrupting friendly behaviors such as grooming, actions referred to as "separating interventions."[40] An especially striking aspect of competition involving coalitions is that males sometimes provide contingent support simultaneously to two males competing for status, a phenomenon termed "allegiance fickleness." Shifting loyalties have been observed in the wild as well as in captivity, leading researchers to infer that males are able to manipulate social relationships to achieve particular goals. In an intriguing case at Mahale, for example, a lower-ranking male alternately supported two rival males whose dominance relationship remained unstable. As a consequence, the higher-ranking males tolerated his attempts to mate with swollen females to whom they would otherwise have prevented him from gaining access.[41]

Toward Females

Females are regular targets of male aggressive behaviors, which range from mild threat gestures to more severe actions such as charging displays, chasing through the canopy or on the ground, and injurious physical attacks. In one year at Gombe, for example, fifty-two male attacks on females were recorded, but only one female attack on a male.[42] In one study each at Taï and Kanyawara, average rates of escalated male aggression toward females were 0.078 and 0.017 acts per hour of female observation time, respectively (or 0.9 and 0.2 acts/ twelve-hour day/male, meaning that if a female is traveling with several males, she may be the target of one or more aggressive acts daily). At Budongo, females received some form of aggression, from mild threat gestures to attacks, at a rate of 0.058 acts/hour (0.7 acts/twelve-hour day/male).[43] Although average rates of escalated aggression by males toward females have been calculated in only two studies, and considerable individual and contextual variation exists within sites, the general pattern of male-to-female dominance behavior appears universal. Moreover, extremely severe attacks by males on community as well as stranger mothers that involve attempted or successful infanticides have been observed at all sites except Bossou (Chapter 8).[44]

BALANCING COMPETITION AND COOPERATION

Although severe and sometimes frequent aggression by males toward both females and other males has been observed at all study sites, it is equally evident that long periods of relative calm occur during which males build close ties through friendly behaviors. The importance to males of maintaining social bonds with other males appears paramount in view of the repeated documentation of the benefits of coalitions in both within- and between-community contexts. Grooming above all appears central to the bonding process, while the importance of repairing relationships potentially threatened by aggressive interactions is revealed by the consistent occurrence of post-conflict reconciliation behaviors (Chapter 1). Although controlled studies of reconciliation are difficult to conduct in the wild, reconciliation rates between males generally appear to be higher than those between females.[45]

Males engage in extremely violent aggression more often than females, a pattern observed in other primate species as well. This is probably because females have more to lose and less to gain from any particular altercation. Female chimpanzees compete for access to nutritional resources. Long-term access to the best feeding areas is clearly important to their reproductive success, but is perhaps best established through repeated, low-level conflicts rather than through physically risky fights over a specific food patch. Moreover, females are often in the company of young offspring, who may be particularly vulnerable during incidents of severe aggression. By contrast, the consequence of successful male aggression can be immediate and potentially ongoing access to fertile females, ephemeral "resources" that offer major fitness benefits if monopolized.[46] The following chapter explores in closer detail how patterns of aggression influence male sexual access and how female sexual behavior reflects the conflict between male and female reproductive strategies.

7

Sexual Behavior: Conflicting Strategies of Males and Females

OVERVIEW

Wild chimpanzee sexual behavior is remarkable in several respects and offers a crucial window into the dynamics of chimpanzee social life generally. In the two- to three-week period leading up to ovulation within the roughly thirty-six-day menstrual cycle, a female's anogenital area swells dramatically, an estrogen-induced phenomenon resulting in a bulbous, bright pink protuberance 900 to 1,400 cubic centimeters in size (e.g., twice the size of a grapefruit) (Plate 7).[1] Estrous swellings are thus highly visible signals of a female's approximate reproductive state. They attract males of all ages, and the majority of sexual behavior in chimpanzees occurs when females are maximally swollen. Since females do not cycle for several years after giving birth to an infant that survives to weaning,[2] competition between males to mate with females that have resumed cycling is intense. Party sizes increase when swollen females are present, bringing many competing males together into what often become highly charged social contexts.[3] Observers have documented a variety of tactics used by males to gain mating access to females at these times.

Mature female chimpanzees, for their part, typically mate an extraordinary number of times when they are in estrus. Swollen females mate with most or all of the males in their community, including adolescents, juveniles, and even infants.[4] Hourly copulation rates for females have been calculated to be between less than one to nearly five per hour at five different field sites, with females mating more often when more males are present. In a study at Ngogo spanning six years, individual females mated with between twenty-eight and thirty-eight different males when fully swollen. In a notable case, one female was observed to copulate sixty-five times with eighteen males in one day. A typical estrous period thus includes hundreds of matings – the record number observed at Ngogo for a single estrous period was 439.[5] Astoundingly, since females usually experience numerous estrous cycles before fertilization, estimates of the average number of times a female copulates per conception range from more than 400 to several thousand.[6]

Despite their high copulation rates with large numbers of males, when and with whom females mate is not random. As noted, sexual activity occurs primarily when females are maximally swollen. However, partners and rates vary over the course of the swelling phase of the menstrual cycle. Matings with infants, juveniles, and adolescents are more common during the initial, roughly one-week period of inflation (called "tumescence").[7] Based on hormonal analyses, ovulation is most likely to occur toward the end of maximal swelling, and thus copulations early in the swelling phase are not likely to result in conceptions.[8] Females mate predominantly with adult males during maximal swelling, the duration of which averages between nine and fourteen days. Most copulations overall occur during this time, with competition between adult males intensifying toward the middle of the phase.[9] It is during this interval, in particular, that alternative mating patterns emerge. Matings subsequently become less frequent as swellings decrease ("detumescence") and are rare when females are "flat."

MATING PATTERNS

Males reveal sexual arousal with a variety of "sexual-invite" displays, most typically by sitting with their thighs spread and penis erect, combined with one of a number of characteristic gestures such as branch shaking or extending an arm toward a swollen female.[10] An erect chimpanzee penis is bright pink and comparatively long. It is thus highly visible against the pale skin of the male's thighs and abdomen. Male displays are rarely ignored by females, who subsequently approach to mate. Males may also approach females, and females usually mate with them when they do.[11] Whether approached by the male or the one to approach, females present their hindquarters to the male in a crouching position. Males either mate from a squatting position with minimal physical contact or mount females from behind. Copulations can take place either on the ground or in the trees, intromission is brief, and females frequently dart away after it is completed.[12] Males often emit a characteristic panting vocalization during copulation, while females sometimes produce a high-pitched "copulation scream" (Chapter 10). It is unclear whether orgasm occurs.

Three different mating patterns have been reported at all study sites: "opportunistic," "possessive," and "consortship."[13] A number of variables influence which type occurs for a particular estrous cycle for a given female, and understanding their interplay remains an active area of field research. In general, it seems evident that the interactions of males and swollen females reflect a degree of conflict between the sexes. On the one hand, a male obviously achieves the greatest genetic benefit if it is his sperm that fertilizes a female. Not surprisingly, individual males often appear motivated to gain exclusive access to fertile females. At the same time, however, females appear motivated to mate with as many males as possible. The outcome of this conflict of interests, so to speak, then depends on a variety of factors, including the availability of supportive alliance partners, the relative number of males and females in communities, the stability of male dominance hierarchies, and the number of simultaneously swollen females.

Opportunistic Mating

Opportunistic mating refers to sexual activity in group contexts in which a swollen female copulates with more than one male without direct interference from other individuals. From 70 to 95 percent of copulations occur in opportunistic contexts, with females often mating with many different males in rapid succession.[14] This seemingly permissive mating atmosphere, however, is not devoid of rivalry. A considerable amount of male aggression may occur in groups containing swollen females, so that it is difficult to evaluate subtler forms of male–male competition.[15] In addition, male chimpanzees are capable of many copulations in succession with intervals of as little as five minutes.[16] They have exceptionally large testes, probably reflecting an adaptation to produce large amounts of sperm per ejaculate, which would be advantageous in sperm competition within a given female and/or in permitting multiple matings with many females that are swollen simultaneously.[17]

Possessive Mating

Possessive or "restrictive" mating refers to efforts by a male, or by two or more males acting in concert, to maintain exclusive sexual access to a swollen female, especially toward the end of maximal swelling when females are most likely to conceive.[18] Possessive males attempt to maintain close proximity to swollen females by following and sitting near them or by inducing the females to follow them by staring at them persistently. Possessive males also aggressively interrupt copulations by other males, charging either or both mating individuals. Restrictive mating situations can last anywhere from an hour to many days.[19] Higher-ranking males, and especially the alpha male, are the most successful at carrying out this strategy.[20] Possessive mating efforts by high-ranking males are concentrated on older, parous females (i.e., females that have already borne offspring), consistent with the general preference of these males for females that have demonstrated their fecundity.[21]

Consortship Mating

Occasionally a swollen female will travel alone with a male away from other group members, a sojourn referred to as a "consortship." Males attempt to initiate consortships in mixed-sex groups by directing a courtship signal toward a female and then walking away from the group while intermittently pausing to stare back at her until she follows. Females sometimes appear reluctant to accompany males, and those that fail to follow may be attacked.[22] Meanwhile, other males may intervene aggressively to prevent a male from leading a female away. If a male succeeds in establishing a consortship, perhaps leading the female to the periphery of the community range, he then maintains exclusive mating access to her for days, weeks, or occasionally even a few months.[23] Relations between the two may be calm once they are separated from other group members. They typically feed, rest, and travel in close proximity and may groom together more than is typical, with the male grooming the female more than the reverse.[24] The pair may also copulate less frequently than in other contexts, reflecting the generally more relaxed atmosphere possible during these excursions.[25]

Relative Success of Different Males in Different Mating Contexts

Conceptions occur in all mating contexts, but evidence to date from several sites indicates that the majority take place in opportunistic settings and the fewest result from consortships.[26] Studies at all sites have shown that higher-ranking males sire more offspring on average than lower-ranking males and that the alpha male sires the most or gets the most copulations (where paternity data are unavailable).[27] Although many males may get the chance to mate in opportunistic contexts, higher-ranking males achieve a disproportionate share of copulations as the female approaches ovulation.[28] Thus the distribution of copulations and conceptions reflects the importance of male competitive behavior and the formation of dominance relationships. This interpretation is reinforced by the observation that higher-ranking males are generally more successful at maintaining exclusive access to swollen females in possessive mating contexts.[29]

Although it seems clear that achieving high rank is reproductively advantageous for males, it is equally evident that lower-ranking males also sire many offspring. There is some evidence that consortship formation provides an alternative mating strategy for lower-ranking males that are unable to compete successfully against more dominant males in group settings: At Gombe, where consortship formation is more common than at other sites, lower-ranking males tend to form them more often than higher-ranking males.[30] Consortships appear to be comparatively rare at Mahale, Kanyawara, Ngogo, and Budongo,[31] and only past, present, and future alpha males at Taï are reported to form them.[32] The success of lower-ranking males at these sites will then depend on their ability to take advantage of mating opportunities in group contexts when females are being ignored by higher-ranking males. Evidence from Taï, for example, indicates that the alpha male achieves a lower proportion of conceptions when there are more male competitors in the community and when more females are swollen simultaneously, contexts in which he is less able to monopolize particular females.[33]

"INDIRECT SEXUAL COERCION"

As noted previously (Chapter 7, "Aggression, Dominance, and Submission"), female chimpanzees are regular victims of male aggression. Although sexual swellings may in some cases stimulate greater tolerance by males, especially toward young immigrant females,[34] males nonetheless attack resident females throughout their sexual cycles. A constellation of evidence from several studies has indicated that such aggression has reproductive payoffs for males. In studies at Kanyawara and Budongo, aggression toward females increased during the periovulatory period (POP, the last few days of maximal swelling, toward the end of which the female ovulates) and was aimed preferentially at older, parous females who are the most fecund.[35] Males then mated most with females toward whom they had directed the most aggression. In addition, females initiated periovulatory copulations preferentially with those males that directed the most aggression toward them throughout their estrous cycles (i.e., when maximally swollen and otherwise).[36] Similarly, over a seventeen-year period at Gombe, males that directed the most aggression toward swollen females had higher copulation rates with those females, and males that were aggressive toward cycling females when they were both swollen and nonswollen were more likely to sire offspring.[37] Since it is *extremely* rare for males to force females to mate with them,[38] aggressive behavior by males toward females has therefore been interpreted as "indirect sexual coercion," whereby females are conditioned into a "fearful respect" of particular males with whom they later mate.[39]

By contrast, researchers at Taï have argued that male aggression in this population does not determine with whom a female copulates when she is most fertile. Instead, Taï females are proposed to exhibit a mixed strategy in which they mate more randomly with males in the early stages of estrus when they are not fertile, but then exercise a choice for more preferred males as they near ovulation.[40] These conclusions are based on two sets of observations. First, it is comparatively rare for male chimpanzees to use aggression in sexual contexts,[41] and male aggression following a female's reluctance to mate does not increase the probability of actually gaining a copulation.[42] Second, one study indicated that swollen females on average appeared to approach some males more than others to mate, and to ignore or avoid some soliciting males more than others.[43] However, this study was not designed to test the possibility that apparent preferences by females around the time of ovulation are a consequence of fear of attack by males that have previously been aggressive toward them.[44] The mixed strategy hypothesis therefore remains speculative.

THE ADAPTIVE SIGNIFICANCE OF FEMALE SEXUAL BEHAVIOR

Whereas aggressive behavior by chimpanzee males appears to offer clear reproductive payoffs, the adaptive significance of female sexual propensities is more puzzling. Two aspects in particular remain to be explained definitively: why females are motivated to mate with many different males (as opposed to being choosy for the "best" male) and why females copulate so many times in each estrous cycle. Both of these characteristics appear to be in conflict with individual male reproductive interests, and although males can sometimes curb female tendencies through possessive or consortship mating, it is evident that there has been strong selective pressure for females to seek out many partners and to copulate at exceptionally high rates – despite receiving considerable aggression from males and enduring persistent feeding interference.[45]

Why Do Females Mate with So Many Males?

Severe intraspecific aggression, which is sometimes lethal, has been observed at all chimpanzee study sites. Although lethal aggression is most frequently directed at members of neighboring communities, males do sometimes target group mates, including infants (Chapter 8). However, if a female copulates with all males in her community and paternity is consequently unclear, then a male that attacks her infant risks attacking his own offspring. Thus, paternity confusion would in theory benefit a female by increasing the potential cost to males of committing infanticide, selecting over time for the propensity in females to mate with many males, and selecting against infanticidal behavior in males.

This "sexual selection" hypothesis, however, is difficult to test empirically.[46] Comparison of the long-term reproductive success of female chimpanzees that vary in this behavior would provide decisive data, but such data are as yet unavailable. Likewise, it would be important to compare the long-term reproductive success of infanticidal versus non-infanticidal males. However, intracommunity infanticide is comparatively rare in chimpanzees, the identity of attackers is not always known, and the collection of long-term data on male reproductive success has only recently become possible with the advent of genetic screening techniques. Therefore, though plausible and widely invoked, the hypothesis remains untested. Other suggested social benefits of mating with many males remain speculative as well – for example, offering females the opportunity to interact with males and subsequently exercise mate preferences[47] or form supportive relationships with them.[48]

Why Do Females Copulate So Many Times?

Although mating with many males might benefit females by inhibiting future infanticide attempts, it is less clear why females copulate so frequently, since a few well-timed copulations with each male could in theory both guarantee conception and confuse paternity. A likely explanation is that high copulation rates again reflect the evolutionary outcome of the conflict between male and female reproductive strategies. On the one hand, estrous swellings provide a conspicuous and highly attractive cue of a female's reproductive state, thus inviting the attentions of all males in the group. If they are indeed "conditioned" to mate with aggressive males, females, especially older parous ones, will be "obliged" to engage in increasing numbers of copulations as more males are present.[49] At the same time, males have the potential to, and sometimes do, use aggression to limit the number of partners females succeed in mating with, especially around the time of ovulation. Since males incur energetic costs by attempting to constrain female mating,[50] high copulation rates by females could also make it more expensive for males to restrict their mating efforts over the entire course of the estrous cycle. Females would consequently have a better chance of distributing enough matings across all males, including those seen less frequently, to confuse paternity completely.[51]

All in all, it is evident that aggression plays an important role in chimpanzee sexual behavior and competition. Frequent aggression by males, in the form of threats, chases, and attacks, clearly influences how often and with whom a female mates. Among themselves, males use aggression to compete for dominance status, which in turn influences their mating strategies. Coalitionary killing, which occurs both within and between communities, offers another important window into the role of aggression in shaping the lives of both female and male chimpanzees. Its contexts and consequences are the subject of Chapter 8.

8

Coalitionary Lethal Aggression between and within Communities

OVERVIEW

Although occasional deaths resulting from fights between individuals have been observed across a wide range of species, coalitionary lethal aggression is common in only a few mammal taxa. It is comparatively frequent in several species of social carnivores and especially well documented in wolves.[1] With the exception of chimpanzees and humans, it is rare among primates.[2] In those species in which it occurs, coalitionary killing typically involves overwhelming superiority in numbers, such that victims are easily overpowered and attackers are at minimal risk of injury. Accordingly, two prerequisites, both met in chimpanzees, appear necessary for the emergence of coalitionary lethal aggression in the behavioral repertoire of a species: the ready formation of coalitions in general and a pattern of ranging in which groups of individuals with the potential to form coalitions regularly encounter isolated conspecifics.

Killings have been observed at all long-term chimpanzee field sites. Isolated adults and older adolescents are typically overwhelmed by multiple aggressors acting in concert. Victims are often forcibly pinned to the ground during attacks, while attackers inflict internal and external wounds by beating them with their hands, stomping on them with their feet, tearing flesh from limbs, and biting all parts of the body, including the testicles of male victims, which are often thus removed. Attacks tend to be prolonged, sometimes lasting more than twenty minutes, and victims eventually cease to resist, apparently entering a state of shock as attacks continue. Adult deaths from inflicted wounds may occur minutes, hours, or a few days after aggressors depart. Infants and dependent juveniles, by contrast, may be grabbed from their mothers and more quickly killed with bites to the head or abdomen. Infants are often eaten.[3]

As of 2014, 152 killings, of which fifty-eight were directly observed, had been reported from fifteen communities.[4] At Gombe, lethal intraspecific aggression is the second most common cause of mortality, behind disease, accounting for seventeen of eighty-six (19.8 percent) deaths with a known cause as of 2008.[5] Although there is considerable variation across chimpanzee communities, median mortality rates from violent aggression rival those reported for small-scale hunter-gatherer and subsistence farming human societies.[6] Killings have the appearance of being intentional, insofar as attackers engage in behaviors not seen in typical dominance interactions, which are usually limited to charging displays, grappling, slapping, and biting the extremities. By contrast, lethal attacks more closely resemble predatory attacks on large prey.[7] As a consequence of the relative frequency, coalitionary nature, and apparent intentionality of lethal aggression in chimpanzees, a few researchers have hypothesized, though not without controversy, that significant parallels exist between some forms of killing in chimpanzees and humans.[8]

Lethal aggression in chimpanzees is typically analyzed according to whether it occurs between or within communities, and whether the victims are adults and adolescents or dependent offspring. The majority of directly observed and inferred killings so far reported have resulted from conflicts between members of neighboring communities (sixty-two of ninety-nine, or 63 percent).[9] Weaned victims of intercommunity killings are most often males (forty-two of forty-nine, or 86 percent), and their attackers are primarily males. Within-community killing of older individuals is comparatively rare (fourteen cases), and only male victims have so far been reported. Infants and juveniles are the most frequent victims of lethal aggression overall. Unlike lethal aggression toward adults, however, infants are targeted by both males and females, both females as well as males are common victims, and attackers can be from the same or neighboring communities.

Because intraspecific killing in chimpanzees occurs in a wide variety of contexts and involves a range of victims and attackers, it is probable that the evolutionary advantages accruing to killers vary. However, it has been difficult thus far to definitively identify these hypothetical benefits. For example, coalitions of males killing other males seems explicable in terms of mate and resource competition. However, why males would kill females, who offer potential reproductive opportunities, is less clear. Lethal aggression toward infants is even more puzzling. Killing infant males from neighboring communities would, for males, eliminate future competitors, but then why kill infant females, future potential mates? Both intra- and inter-community infanticide by males would increase the reproductive success of killers if they were subsequently able to impregnate the infants' mothers, but evidence for this is scant – and as previously discussed (Chapter 7), intracommunity infanticide also carries the risk of killing one's own offspring. Meanwhile, intracommunity infanticide by females could eliminate future resource competition for themselves and their offspring, or it could simply offer a nutritional opportunity. The following sections consider data relevant to these various possibilities. The role that humans may play in augmenting rates of lethal aggression in chimpanzees is considered in the final section.

KILLING ADULTS FROM NEIGHBORING COMMUNITIES

The general context of intercommunity lethal aggression is closely similar across sites and well described. Attacks occur when groups of males traveling or foraging in areas of overlap between neighboring communities, or encroaching some distance into an adjacent territory, encounter comparatively isolated neighbors. Confrontations between male neighbors often involve loud vocal and charging displays followed by the retreat of one group, often before visual contact is made.[10] Where numerical superiority is evident, however, larger groups will attack smaller parties or individuals, and it is likely that boundary areas are often avoided for this reason by individuals foraging alone or in small groups. Differences between study sites in the frequency with which deadly attacks occur then offer opportunities to explore how ecological and demographic variables may affect the likelihood of lethal interactions between communities.

Border Patrols and Lethal Raiding

Researchers observe two distinct patterns of movement in the vicinity of community boundaries. On the one hand, foraging parties may produce loud vocalizations upon approaching the edge of their range and, if they hear no calls in response, continue into an area to feed on seasonally available foods. Smaller groups or individuals may simply move into and feed quietly within these border zones. By contrast, parties of males, sometimes accompanied by females, may travel in compact and exceptionally quiet groups toward and along boundaries, evidently scouting for the presence of neighbors. Individuals in these groups appear alert, often move in single file, feed little, pause regularly to inspect and sniff the ground, and periodically sit quietly to listen, apparently, for sounds of neighbors. Regular occurrences of these "border patrols" have been reported from five study sites, with observations consistent with patrolling behavior reported from a sixth (Plate 8).[11] The frequency of reported patrolling varies from as often as nine times in a month to once a month or less.[12]

In addition to foraging and patrolling in boundary areas, groups of males also travel directly into neighboring territories for substantial distances, sometimes a kilometer or more. Such penetrations into the ranges of other communities are referred to as "deep incursions." As with patrolling in general, deep incursions are more likely to occur when parties contain many males.[13] Individuals behave as they do during patrolling, with the same indications of caution and searching, suggesting that they are hunting for neighbors rather than foraging. This inference is supported by observations in which deep incursions end in killings when isolated neighbors are encountered and attacked.[14] Killing in the context of deep incursions has been dubbed "lethal raiding" and compared to stealthy raiding in small-scale human societies engaged in violent intergroup conflict.[15]

The "Imbalance-of-Power Hypothesis"

Since killing competitors would seem to be the most effective means of guaranteeing access to resources, the question arises as to why it is not more common among animals. An obvious answer is that killing rivals, especially morphologically similar conspecifics, is physically difficult and therefore potentially costly in terms of energy expenditure and risk of injury. These impediments can be removed, however, when attackers greatly outnumber victims. Thus among chimpanzees, males appear to assess the relative numerical strength of neighboring parties before interacting with them. This impression is supported by evidence from a field experiment conducted at Kanyawara, in which only parties containing three or more males responded to playbacks of an unfamiliar "pant hoot" vocalization (see also Chapter 10, endnote 40).[16] Moreover, victims of lethal attacks are typically completely immobilized, and there have been no reports of attackers sustaining serious injuries. The explanation for why groups of chimpanzees kill other chimpanzees, as opposed to simply engaging in non-lethal forms of aggression, may therefore simply be because they easily can. This proposition forms the basis of the "imbalance-of-power hypothesis," which posits that coalitions of group members kill conspecifics because the costs are trivial.[17]

If the imbalance-of-power hypothesis is correct, then differences between communities in rates of intergroup lethal aggression may be largely explicable in terms of differences in the vulnerability of group members and in the number of males available to form coalitions. For example, members of the primary study community at Taï (North group) have not been victims or perpetrators of lethal intercommunity attacks. On the one hand, average party size in this community is high, and individuals therefore spend comparatively less time foraging alone and vulnerable to attack.[18] At the same time, North group suffered a precipitous decline in male numbers in the first decade of research after habituation was completed in 1982, culminating in just three males by 1995. Opportunities for safely attacking isolated neighbors would be rare with so few males. By contrast, two lethal attacks by males from the South study group at Taï have now been documented. At the time of the observations, South group included six to eight adult and adolescent males. The victim in the 2002 attack was an infant from the Middle study group. The victim in the 2005 attack was an adult male from the semi-habituated East study group, who was apparently foraging alone before being overwhelmed by four adult and two adolescent males.[19]

Lethal Aggression and Territorial Expansion

The importance of intercommunity competitive strength for gaining access to food resources was first indicated by studies of neighboring groups at Mahale whose territories overlapped. Prior to observations of intergroup lethal aggression, observers there documented seasonal ranging patterns in which the smaller K-group was consistently displaced to the north as M-group migrated northward to occupy what was ordinarily an overlap zone between the two communities.[20] Later reports of males from larger communities killing many members of neighboring groups emphasized the importance of community size, and number of community males in particular, for maintaining access to desirable ranging areas. At three different sites – Gombe, Mahale, and Ngogo – repeated killings resulted in substantial territorial expansion that lasted for years. Since lethal incidents often occurred in the absence of immediate resource competition, such as a food patch subsequently exploited, the benefit of killing has been interpreted to inhere in the long-term advantage of reducing the number of neighboring male competitors.[21]

Community Extermination at Gombe[22]

In the first decade of research at Gombe, researchers noted that the nineteen adult and adolescent males of the main study community (Kasekela) tended to associate preferentially in two separate subgroups. By 1972, for reasons that were unclear, the smaller subgroup splintered off along with three females and restricted its movements to the southern part of the original community range. Interactions between males from the two groups were henceforth hostile, a permanent division that remains the only community fission event ever documented in chimpanzees. The southern group, known as the Kahama community, included four prime-aged adult males, one of whom was crippled from polio; two old males; one adolescent male; and three females. Six prime males, two old males, and two adolescent males stayed in the Kasekela community, which occupied a territory encompassing approximately 15 km².

Between 1974 and 1977, researchers witnessed prolonged attacks by groups of three to seven Kasekela males on each of four isolated Kahama males. These attacks resulted in severe wounding and the subsequent disappearance of the victims, all of whom were presumed to have died as a result of their injuries. The heavily battered body of a fifth Kahama male was discovered close to where local fishermen had heard sounds of fighting and shortly after had observed five chimpanzees pass by their huts. The crippled male and one of the two old males disappeared during this time as well. Researchers also observed a series of severe attacks over the course of twelve months on one of the three Kahama females; this victim died five days after the final episode. The remaining two females disappeared, having either met similar fates or transferred to other communities. These events were particularly striking because the attackers and victims had all previously exhibited affiliative social relationships with one another, the males in particular having spent their entire lives together. Over the course of the conflict, the Kasekela range increased to 17 km², and by 1977 it completely encompassed the previous core area of the Kahama community.

Community Extinction at Mahale[23]

The original study community at Mahale, K-group, consisted of twenty-one individuals in 1967, including six adult males and one adolescent male who came into adulthood by 1971. The eldest of these males disappeared in 1969, having presumably died of old age, reducing the number of males to six. Between 1970 and 1983, all of the remaining males, none of whom appeared old enough to die of natural causes, disappeared one by one. When the number of males was reduced to just two, the mature females in the group began regularly associating with males from other communities. When just one male remained, most of the mature females transferred permanently to other groups. By 1983, the only individuals remaining in K-group who did not associate with other groups were a lactating mother, her infant, and an old female with her adolescent male offspring.

Researchers at Mahale concluded that the disappearances of K-group males were a consequence of attacks by M-group males. There were several grounds for this interpretation. Seasonal migration patterns had previously indicated that M-group was dominant to K-group, and as of 1979 M-group numbered about 100 individuals and included eleven adult and nine adolescent males.[24] Interactions between males of the two groups during encounters in their overlap zones were always aggressive, dominated by charging displays and loud vocalizations, with occasional fighting and wounding.[25] Deep incursions by M-group males into K-group's core area were also observed during this time, and in at least one case M-group males are suspected to have killed and eaten a K-group infant.[26] Finally, after the disappearance of the K-group males, M-group expanded its territory to include K-group's former core area.

Twenty-One Killings and Territorial Expansion at Ngogo[27]

Researchers at Ngogo documented the violent deaths of twenty-one individuals from neighboring chimpanzee communities over a period of ten years, from 1999 to 2008. In thirteen cases, twelve of which occurred during border patrols, death ensued shortly after observed coalitionary attacks by Ngogo males. In five cases, a coalitionary attack was observed that resulted in severe wounding similar to those cases in which the death was observed. Researchers inferred that coalitionary attacks had occurred in three remaining deaths. In one of these, chimpanzee hair and bone were found in the feces of a Ngogo male that on the previous day was part of a group of males seen attacking a neighbor female. The researchers inferred that the female's infant had been killed and eaten. In a second case, researchers followed males into their neighbor's territory, where they found another party of Ngogo males consuming a dead infant. In a final incident, researchers found the severely wounded corpse of a neighbor male in an area where the previous week they had lost contact with a party of patrolling Ngogo males they were following.

By 2009, the Ngogo chimpanzees had augmented their community range by 6.4 km², an increase of roughly 22 percent over their former 28.7 km². The expansion occurred in the area northwest of their former range. Eleven of the fatal attacks had occurred within the newly annexed area, and three more took place close by. Over the ten years during which these killings were documented, the Ngogo community was composed of 145 to 160 individuals, with twenty-five to thirty adult males and eleven to twenty-two adolescent males. The unusually large size of the community, including its exceptional number of males, likely explains the frequency and potency with which it was able to attack its neighbors.

Possible Benefits of Territorial Expansion

While intergroup hostility between males has the effect of preventing neighbor males from gaining mating access to community females, the precise benefits that males derive from territorial expansion via lethal aggression are less clear. Two general hypotheses have been proposed. First, territorial expansion could result in female recruitment, directly or indirectly. Since females within communities tend to restrict their ranging to smaller, overlapping core areas (Chapter 4), males might directly recruit neighbor females into their community by gaining control of the areas they frequent. Alternatively, females might be recruited indirectly if larger territories with more resources are sufficiently attractive to entice them to transfer. Second, males might benefit in the long term by improving the quantity and/or quality of the community's nutritional resource base, both for themselves and for community females and their offspring.

There is little evidence to support the female recruitment hypothesis. Female transfer following lethal intercommunity aggression has only been reported in the cases in which most or all of the smaller community's males were killed or disappeared.[28] In contrast, analysis of Gombe data gathered from 1975 to 1992 showed that the number of females did not increase when the community range expanded nor decrease when it contracted.[29] Instead, females shifted their cores areas to remain within community borders as the community range fluctuated. Since cases of community extermination appear to be very rare, female recruitment is therefore unlikely to be the ultimate explanation for the otherwise comparatively frequent occurrence of lethal aggression across sites. The hypothesis is additionally weakened by observations at all sites of severe attacks on stranger females, which would seem contrary to a strategy of recruitment (but see discussion of infanticide later).[30]

By contrast, two lines of evidence from Gombe suggest that males may benefit reproductively from territorial expansion by enhancing their community's resource base. First, as discussed in Chapter 4, data from all long-term studies indicates that the average size of chimpanzee foraging parties increases when the availability of ripe fruits increases. Between 1975 and 1992 at Gombe, the average size of foraging parties, and in particular of mixed-sex parties, increased with increased territory size, suggesting improved food abundance. This was also accompanied by an increase in the number of days that males encountered swollen females. Second, the fecundity of Gombe females improved with more territory. Studies of provisioned and captive primates demonstrate that female reproductive rates and infant survivorship increase when females are better fed and heavier.[31] Between 1973 and 2000, Gombe females weighed more when the community range was larger. Between 1975 and 1992, likewise, females had shorter interbirth intervals when the territory size was greater.[32] Thus, by expanding their range, Gombe males appear to have increased their reproductive potential by improving the community's access to preferred foods.

KILLING ADULT GROUP MEMBERS

Although much less common than violent intercommunity aggression, severe coalitionary attacks on group members that result in death also occur occasionally. As of 2014, fifteen intracommunity killings of weaned individuals, of which twelve were adults, were reported from six field sites.[33] Of the fifteen victims, thirteen were males of varying ranks, of which one was an alpha and another a former alpha. Two individuals from Gombe died with unhealed wounds some months after severe attacks: one, an old adult male, disappeared two months after receiving a deep facial wound; the other, a twelve-year-old female, disappeared during a pneumonia epidemic six months after receiving a severe groin injury.[34] An additional attack on an alpha male at Gombe probably would have resulted in death had researchers not intervened and administered antibiotics to the victim.[35] At Mahale, two adult males disappeared for twelve and fifty days, respectively, after surviving severe coalitionary attacks. The first had been the alpha male. He lost his alpha status upon return, regained it fourteen months later, and then disappeared completely three months after that.[36] The second victim was middle ranking and reported to have had an unstable position in the male social hierarchy, indicated by his failure to exhibit the typical gestures of submission to higher-ranking males.[37]

The comparative rarity of lethal intracommunity attacks on adults by coalitions of males, combined with a lack of information about many of them, makes it difficult to explain their causes and benefits. Nevertheless, the great majority of victims are males, and at least some of the incidents appear to be a consequence of direct or indirect sexual competition. Evidence from all study sites indicates that male rank is correlated with reproductive success (Chapter 7). Three severe attacks, one of which was lethal, targeted alpha males in the context of dominance status challenges.[38] Other cases involved either lower-ranking males in the process of ascending the status hierarchy or very old males.[39] While the formation of coalitions in the context of male status striving is commonplace (Chapter 6), lethal aggression presumably is rare because the costs of reducing male numbers and losing potential allies, who are critical for support in intercommunity conflicts, outweigh the benefits of increased rank. It is perhaps for the same reason that killing prime-aged males, who patrol most and are key participants in intercommunity aggression, is uncommon. Two lethal attacks on community females at Budongo remain unexplained.

KILLING INFANTS

The first reported chimpanzee infanticide was observed at Budongo in 1967.[40] Since then eighty-five cases have been observed or inferred, making infanticide the most common form of lethal aggression in chimpanzees. It has been reported from all long-term field sites and occurs both within and between communities.[41] Since both males and females are known to violently attack adult females, it is possible that some infants are unintended victims. However, repeated observations indicate that infants are regularly the focus of attacks. In the most dramatic cases, mothers are physically restrained by one or more attackers as accomplices reach for and seize infants, with the mother subsequently released as the attackers make off with the victim.[42] In many cases infants are eaten after being killed, by both female and male attackers.

It is unlikely that there is a single functional explanation for infanticide since the contexts in which it occurs are variable. At the very least, it is probable that the potential long-term benefits for male and female killers differ, as well as the reasons for killing the infants of neighbor versus resident females.

Infanticide by Male Attackers

Intercommunity Infanticide

Males could benefit reproductively by killing the infants of neighboring females if the females were thereby induced to transfer into the killer's community. This explanation for male infanticidal behavior is founded on sexual selection theory, which accounts for the evolution of behaviors that benefit individuals by enhancing their access to reproductive opportunities (Chapter 7, endnote 46). As discussed previously, however, adult females appear to transfer only after their community males have disappeared, not when their infants have been killed. Alternatively, intercommunity infanticide could benefit males if it reduced the future competitive strength of the neighboring community. This hypothesis would be supported if males primarily killed male infants, sparing female infants that would ultimately transfer and therefore represent potential mating partners.[43] However, it has become evident that both male and female infants are targeted.[44] Accordingly, a third possibility is that intercommunity infanticide facilitates range expansion, inducing neighbor females and associated males to forage away from border areas.[45] The reproductive benefits for males of range expansion, as discussed previously, include more frequent interactions with resident females and improved female fecundity.

Intracommunity Infanticide

The potential long-term costs to males of killing infants within their own community would seem to be substantial, and the behavior therefore remains difficult to explain. As already discussed (Chapter 7), since female chimpanzees mate with many males when they are in estrus, infanticidal males would risk killing their own offspring.[46] And even if males could distinguish related from unrelated infants, the costs of killing would still seem to outweigh the benefits. For example, although killing male infants would reduce the number of future competitors for status as well as nutritional resources, it would also reduce the number of potential future allies available, either for the male or his male offspring, to defend the community range. Killing female infants might similarly reduce future resource competition, but it would also reduce the availability of future mates overall by eliminating female breeders in the population. Finally, killing unrelated infants would benefit males if they were subsequently able to mate restrictively with the victim's mother, as occurs in some other primate species.[47] However, there are no reports of male killers maintaining exclusive mating access to mothers after infanticide.

On the other hand, researchers at Mahale have suggested that within-group infanticide might indirectly increase male mating opportunities by inducing females to mate more restrictively with community males in general. In support of this idea are several cases in which the mother of the victim occupied a core area on the periphery of the community range and had been frequently absent from the community when her infant was conceived. Subsequent to their infants being killed, these mothers were observed to associate more regularly with the community's adult males. [48] In other cases of infanticide at Mahale, the mothers had mated predominantly with lower-ranking or adolescent males around the time of conception. The proposed benefit to the adult male killers in these cases was to induce the females to mate more restrictively with higher-ranking males.[49]

Infanticide by Female Attackers

Of a total of forty-five intracommunity infanticides by females so far observed, inferred, or suspected, the attackers were identified in twenty-three cases. Of these, nearly half (ten cases) were by coalitions of adult females at three different field sites.[50] In the cases reported from Gombe and Taï, the female attackers ate the victims. Apart from these successful attacks, there are several reports of failed attempts by coalitions of females to grab infants during episodes of intense aggression toward mother–infant pairs.[51] There are three additional reports of females participating with males in successful intracommunity infanticides, and presumably some of the twenty-two cases in which the attackers were never identified involved female attackers.

The frequent consumption of infant victims suggests the straightforward adaptive hypothesis that females benefit nutrition-ally by killing and eating infants.[52] Nevertheless, a nutrition-by-predation hypothesis fails to explain the many infanticide cases in which the victims are not eaten. Consequently, a more general adap-tive hypothesis for infant killing by females is that it reduces resource competition. As discussed in Chapter 5, females are extremely aggres-sive toward new immigrants and compete with other resident females for access to the best feeding areas. Killing the infants of other resi-dents would reduce feeding competition in three ways.[53] First, the nutritional requirements of primate females increase by as much as 50 percent when they are lactating.[54] The feeding needs of the infant victim's mother would therefore decline with the loss of her infant. Second, compared to females with offspring, females in estrus tend to travel more widely in association with males. Thus, the vic-tim's mother would be likely to spend less time directly competing with an infanticidal mother that is foraging in a more narrowly restricted area of the community range. And third, a future competi-tor of the attacker and her offspring would be eliminated.

IS CHIMPANZEE LETHAL AGGRESSION "NATURAL"?

Early reports of lethal aggression from Gombe and Mahale prompted the suggestion that the killings were a direct consequence of provision-ing the study animals and abnormally intensifying resource competition.[55] However, subsequent observations of killings at sites lacking provisioning indicated that intraspecific lethal aggression is a normal component of the evolved behavioral repertoire of chimpan-zees. In support of this conclusion, a recent effort to quantify the relationship between human activities and killing rates failed to find a direct link to any of several possible types of human influence, including habitat disturbance and disease exposure. This study instead concluded that increased killing was associated with higher population density and the number of males in communities.[56]

Nevertheless, primate researchers elsewhere have explored how long-term contact with humans can influence primate popula-tions in subtle ways.[57] Thus the possibility remains that human popu-lation pressure and resource extraction have, at the very least, affected rates of deadly violence among chimpanzees.[58] For example, adaptive explanations for the most frequent forms of lethal aggression in chimpanzees ultimately focus on competition for space. Male attackers are hypothesized to gain better mating opportunities and achieve higher reproductive success by expanding their territory subsequent to lethal intercommunity interactions. Females, who may benefit from killing infants by reducing feeding competition in general, also profit from territorial expansion by gaining access to better feeding areas. This implies, therefore, that variation in the frequency of lethal aggression within and between sites may result at least partly from differences in the size and quality of the ranges that communities defend. In view of the continued human population growth, agricultural development, and resource extraction occurring across the African rain forest belt, it will therefore be important to examine whether human encroachment augments levels of chimpanzee lethal aggression. This could result, for example, if habitat disturbance caused range compression and more frequent contacts between neighboring chimpanzee communities.

Additionally, a potential human influence on chimpanzee territorial behavior that has not been examined empirically concerns the possible impact that researchers themselves have on intercommunity interactions. Unhabituated chimpanzees are extremely fearful of humans and flee when they see them. Consequently, researchers closely following a group of habituated chimpanzees that come into contact with unhabituated neighbors could affect the outcome of the encounter in at least two ways. First, they could simply scare the unhabituated animals into flight, facilitating territorial encroachment by the study animals. If contact with humans also inhibited the neighboring chimpanzees from frequenting the area in the future, a more long-term boundary shift could ensue. Second, fleeing animals could become separated from their group and more vulnerable to attack by the patrolling animals. Although no intercommunity lethal attack has been attributed to such a circumstance, the possibility is nonetheless real. These considerations suggest the need for systematic analyses of the impact of human observers on intercommunity interactions.[59]

Coalitionary lethal aggression thus appears to be a normal component of the chimpanzee behavioral repertoire, though the frequency of its occurrence may be influenced by human activities. Whether it is relevant for theorizing about the evolution of lethal violence in humans is more controversial. Chimpanzee coalitionary killing obviously lacks the cultural and institutional complexities of human warfare, but there may nonetheless be functional parallels between it and lethal raiding in small-scale human societies that could shed light on the evolution of violent intergroup conflict in human ancestors (endnote 8). At the very least, however, it highlights the complex mix of competition and cooperation that characterizes adult male chimpanzee sociality. Males must continuously navigate tense dominance relationships among each other while also engaging periodically in collective and oftentimes violent territorial behavior. This, in turn, has a profound impact on females, necessarily influencing their ranging and sexual behavior. And as Chapter 9 describes, this mix of collective action and competition also characterizes another remarkable feature of chimpanzee life in most communities, namely, group hunting and meat sharing.

9

Hunting, Eating, and Sharing Meat

OVERVIEW

Wild chimpanzees hunt and consume a wide variety of small and medium-sized mammals as well as the occasional bird, small fish, and even snake (Table 9.1). Hunting is observed regularly at five of the seven long-term field sites, has been documented in many other populations, and probably occurs wherever prey species are available.[1] Average frequencies at sites where it is common range between four and ten hunts per month, with considerable monthly and annual variation within and between communities.[2] Both male and female chimpanzees hunt, but males hunt much more often and are responsible for the majority of kills everywhere.[3] Primates are their most frequent prey, including individuals from at least sixteen monkey and five prosimian species (not including regional forms of red colobus and black and white colobus monkeys) as well as infant chimpanzees and, rarely, infant humans (Chapter 8).

Although meat constitutes only a small fraction of the chimpanzee diet, accounting for less than 5 percent of their feeding time on average,[4] it is highly prized, as judged by the social excitement that accompanies episodes of prey capture and consumption. Unlike ripe fruits, which are the mainstay of chimpanzee diets at all study sites (Chapter 1), most prey are difficult to capture and are therefore usually hunted by groups of animals. Once a prey individual is killed, however, its carcass can be monopolized, inviting often intense competition for access to meat, with hunters as well as non-hunters performing charging displays, harassing possessors, avoiding aggressors, and vocalizing. An individual who makes a kill may succeed in retaining possession of the quarry, may have some or all of it taken from him or her, or may share parts of it with others (Plate 9).

Table 9.1 *Prey species of chimpanzees at seven long-term (forest) and two shorter-term (savanna) field sites*[1]

		Field site								
	Tanzania		Uganda			Guinea	Ivory Coast	Senegal		
Species	Gombe[2]	Mahale[3]	Kanyawara[4]	Ngogo[5]	Budongo[6]	Bossou[7]	Tai[8]	Mt. Assirik[9]	Fongoli[10]	Common name
Monkeys										
Cercopithecus mitis	•	•	•	•	•					Blue monkey
C. ascanius	•	•	•	•	••					Redtail monkey
C. aethiops		•							•	Vervet
C. l'hoesti				•						l'Hoest's monkey
C. diana							•			Diana monkey
C. petaurista							••			Lesser spot-nosed guenon
C. mona							••			Mona monkey
C. pogonias										Crowned guenon
Cercocebus atys							•			Sooty mangebey
Colobus spp.			•	•	•		•			Black and white colobus
Erythrocebus patas									•	Patas monkey
Lophocebus albigena			•	•						Gray-cheeked mangebey
Papio anubis	•	••		•						Olive baboon
Papio papio									•	Guinea baboon

Table 9.1 (cont.)

	Field site									
	Tanzania		Uganda			Guinea	Ivory Coast	Senegal		
Species	Gombe	Mahale	Kanyawara	Ngogo	Budongo	Bossou	Taï	Mt. Assirik	Fongoli	Common name
Procolobus spp.	•	•	•	•			•			Red colobus
Procolobus verus		•					•			Olive colobus
Strepsirrhines										
Galago senegalensis	•	•						•	•	Lesser bushbaby
G. demidovii				•						Demidoff's bushbaby
G. crassicaudatus		•								Thick-tailed bushbaby
G. alleni										Allen's bushbaby
Perodicticus potto							•	•		Potto
Hominoids										
Pan troglodytes	•	•		•	•					Chimpanzee
Homo sapiens	•									Human (infant)
Ungulates										
Tragelaphus scriptus	•	•		•					•	Bushbusk
Nesotragus moschatus				•						Suni
Potamochoerus porcus	•	•		•						Bushpig

Species	Common name						
Philantomba monticola	Blue duiker		•			•	
Cephalophus spp.	Red duiker		•		•		
Phacochoerus africanus	Warthog			•	•		
Insectivores							
Rhynchocyon cernei	Elephant shrew		•			•	
Rodents							
Funisciurus sp.	Squirrel	•	•				
Protoxerus or *Heliosciurus*	Squirrel						
Cricetomys emini	Pouched rat		•				
Thryonomys swinderianus	Cane rat		•				
Rat or mouse sp.		•				•	
Carnivores							
Mongoose sp.	Mongoose		•				
Mungos mungo	Banded mongoose						•
Viverra civetta	African civet		•				

Table 9.1 (*cont.*)

	Field site									
	Tanzania		Uganda			Guinea	Ivory Coast	Senegal		
Species	Gombe	Mahale	Kanyawara	Ngogo	Budongo	Bossou	Taï	Mt. Assirik	Fongoli	Common name
Myracoidea										
Heterohyrax brucei		•								Rock hyrax
Pangolin						•				
Small fish, shrimp	•					•				
Bird spp.	•			•		•	••			
Snake								•		

[1] Data based on direct observation or fecal remains.
[2] Gombe: •, data compiled in Goodall (1986: Table 11.1).
[3] Mahale: •, data for 1966–1995 compiled in Uehara (1997); ••, data for 1995–2009 compiled in Hosaka (2015a).
[4] Kanyawara: •, Gilby et al. (2017).
[5] Ngogo: •, Watts and Mitani (2015: Table II).
[6] Budongo: •, Suzuki (1971); ••, data compiled in Reynolds (2005: Table 4.6).
[7] Bossou: •, Sugiyama and Koman (1987: Table 2).
[8] Taï N: •, Boesch and Boesch (1989: Table 3); Boesch and Boesch-Achermann (2000: Table 8.1). Note that Taï does not have a regular fecal sampling protocol.
[9] Mt. Assirik: •, McGrew et al. (1988).
[10] Fongoli: •, Pruetz et al. (2015).

High levels of arousal and competition during and following hunts suggest that meat is nutritionally important, presumably because it is a concentrated and easily digested source of protein and fat and provides a valuable source of micronutrients.[5] Nevertheless, the fundamentally social nature of predation by chimpanzees raises a range of questions related to cooperation, exchange, and predator–prey dynamics and has consequently stimulated intensive research at the five long-term sites where it is most frequently observed. Several important issues remain unresolved and are the focus of ongoing investigations. For example, what factors influence the "decision" to hunt? Do chimpanzees cooperate for mutual benefit during hunts? Why do individuals share meat? Why are some individuals more successful at acquiring meat than others? And what are the conservation implications of high levels of chimpanzee predation on other endangered species?

HUNTING PATTERNS, PREFERRED PREY, AND SUCCESS RATES

Hunting techniques vary depending on the type of prey species targeted.[6] A minority of hunts are opportunistic and swift, for example, when a chimpanzee encounters a bushbuck fawn by chance and kills it immediately. Similarly, chance encounters with bushpigs, detected in the course of ordinary foraging and travel when rustling sounds are heard in the undergrowth, offer occasional opportunities to seize piglets. Predation on numerous small mammals and birds appears likewise opportunistic. By contrast, hunts of group-living arboreal monkeys, which account for the majority of kills, can involve one to two hours of persistent watching, stalking, and chasing by large groups of chimpanzees. Group hunts at some study sites occur subsequent to apparently chance detection of monkey groups initially either seen or heard from a distance. In at least two communities, however, chimpanzees appear to actively search for monkey groups in a fashion similar to boundary patrols.[7]

Although lone hunters can sometimes capture arboreal monkeys, group hunts are far more common. Once a hunting party locates a monkey group, it approaches and monitors it from the ground before attacking. Attacks begin with one or several chimpanzees climbing into the canopy to actively pursue potential prey, causing the monkey group to disperse. Other chimpanzees may remain on the ground and track the monkeys from below. In prolonged hunts, a particular chimpanzee is likely to occupy more than one position relative to fleeing prey, at some times in active pursuit but at others closing off escape routes of panicked animals. Captures therefore occur in a variety of ways. Arboreal hunters may succeed in catching fleeing individuals directly, snatch infants from their mothers, or seize mothers defending their young. Fleeing monkeys may fall and be grabbed on the ground by waiting chimpanzees, or they may be caught by hunters waiting below when attempting to descend trunks.

Chimpanzees often appear more adept at capturing than at killing and consuming their prey. Infants may be quickly killed by a bite to the head or neck, but older and larger prey are repeatedly flailed against branches or the ground, or slowly die while being eaten.[8] Consumption of infants typically begins with the brains, which are extracted with the fingers, either through the foramen magnum at the base of the skull or through a crushed cranial bone. When feeding on older and larger prey, by contrast, the viscera and limbs are often eaten first, leaving the head for last.[9] Chimpanzees tear flesh off their prey with teeth and hands, sometimes chewing the meat together with leaves. Small bones may be chewed and swallowed, along with hair, while bigger bones are gnawed and licked. Consumption is generally slow, taking place over the course of hours, and prey may be entirely or only partially eaten. Partially eaten animals are occasionally retrieved hours later or the following day.

Red colobus monkeys are by far the preferred prey of chimpanzees at all sites where the two species are sympatric (Table 9.2). These arboreal monkeys, which are roughly a quarter of the size of chimpanzees, associate in mixed-sex social groups comprised of less than ten to as many as eighty-two individuals.[10] They initially respond to the presence of chimpanzee hunting parties by giving alarm calls and moving away through the canopy. Once pursuit begins, females, young, and at least some males typically scatter. Some adult males may engage in vigorous defensive behavior, which can include vocal and branch-shaking displays as well as chasing, mobbing, grappling with, and biting their pursuers. Such counterattacks can repel solitary or small groups of chimpanzees, and female and immature monkeys are less likely to be captured if they remain near these defending males.[11]

Despite the sometimes-effective defensive behavior of red colobus males, success rates for group hunts of these animals are high, ranging between 40 and 60 percent of attempts in most communities. Roughly 20 to 40 percent of successful hunting episodes result in multiple kills, including victims of all ages and both sexes, with immatures the most frequent victims. In all communities, male chimpanzees

Table 9.2 *Chimpanzee predation on red colobus monkeys at four field sites over time: Preferences and success*

	Study site							
	Gombe[1,2]	Gombe[3]	Gombe[4,5]	Mahale[6]	Mahale[7]	Taï[8]	Ngogo[9,10]	Ngogo[11]
% of prey that were red colobus (Total # prey)		62.2% (259)	82.0%[5] (429)	55.0% (100)	83.3% (245)	77.8% (81)	88.4%[9] (292)	82.7% (1102)
% of encounters chimps hunted (Total # encounters)	75.3%[1] (85)		72.6%[4] (729)		63.2% (117)	7% (est.)	37.2%[10] (164)	55.6% (633)
% of hunts successful (≥ 1 victim) (Total # hunts)	48.4%[1] (64)	40.7% (216)	51.6%[4] (529)		59.5% (74)	54.2% (83)	81.7%[9] (82)	85.3% (356)
% of successful hunts with > 1 kill (Total # successful hunts)	19.3%[1] (31)	42.0% (88)	30.2%[5] (215)	22.5% (40)	23.1% (169)	25.5% (55)	86.6%[9] (67)	80.7% (304)
% of victims that were immatures (Total # victims whose relative age could be determined)	65.5%[1] (29)	77.7% (130)	75.9%[5] (429)	70.0% (40)	80.6% (155)	53.4% (58)	73.6%[9] (258)	
% of hunts that were group hunts (Total # hunts)	70.3%[2] (64)			100% (34)	97.3% (74)	92.5% (80)	100%[9] (82)	

Table 9.2 (cont.)

	Study site							
	Gombe·	Gombe	Gombe·	Mahale	Mahale	Taï	Ngogo·	Ngogo
% kills made by males			89.3%[5]	80.5%	72.6%	81.6%	98.5%[9]	
(Total # kills in which age/sex of capturer could be determined)			(313)	(41)	(117)	(38)	(261)	

[1] Busse (1977), data for 1973 and 1974.

[2] Busse (1978).

[3] Calculated from Goodall (1986: Tables 11.2–11.4 and p. 272 for percentage immatures) for 1975–1981. Table 11.12 lists twenty-two kills for females from 1974 to 1981, but Table 11.2 does not provide a total kill tabulation for the same time period. Additionally, Goodall (1986: 304) notes that until 1976, females were not regularly the targets of full-day follows. Therefore, no value for percentage kills by males is given here, although Stanford et al. (1994b) say that males accounted for 96 percent of red colobus kills from 1971 to 1981, citing Goodall (1986) and Wrangham and Bergmann Riss (1990), the latter reporting data from 1972 to 1975.

[4] Stanford et al. (1994a), data for 1982–1992.

[5] Stanford et al. (1994b), data for 1982–1991. Gilby et al. (2017: Table 2) also reported that the percentage of prey that were red colobus was 82 percent for both the Kasekela (1976–2013) and Mitumba (2000–2014) communities at Gombe, as well as 85 percent at Kanyawara (1996–2015).

[6] Uehara et al. (1992), for data collected during five study periods between 1983 and 1990. Data for percentage kills by males includes all predations (Uehara et al., 1992: Table 5).

[7] Hosaka et al. (2001), for data collected during seven study periods between 1991 and 1995. Group hunting percentage for 1991 and 1993 only (Hosaka et al., 2001: Table 10). For 1996–2009, Hosaka (2015a) reported a similar percentage of red colobus taken (84.3 percent of 254 kills), and comparatively more kills by adult females (18.9 percent of 95 kills versus 22.2 percent of 117 kills).

[8] Boesch and Boesch (1989: Tables 3, 5, 7, 10, 12; p. 554 for estimate (est.) of percentage of encounters hunted, p. 556 for multiple kill rate). Group hunt percentage is for all hunts; these authors say (p. 558) that the figure for red colobus only is 94 percent, but they provide no sample size.

[9] Watts and Mitani (2002: Tables I, III; Figs. 1, 2; p. 14 for captor age/sex; p. 10 for no solo hunts).

[10] Mitani and Watts (2001: 919) for percentage of encounters that hunted.

[11] Watts and Mitani (2015), data compiled for 1995–2014. Total successful hunts calculated from 85.3 percent of 356 hunts (p. 736).

make the most kills, although females have been credited with one-third of captures in some studies. With its unusually large number of males, the Ngogo chimpanzees have exceptionally high success and multiple kill rates (81.7 and 86.6 percent, respectively). The highest recorded number of kills for a single successful group hunt of red colobus monkeys at Ngogo, remarkably, is thirteen individuals.[12] By contrast, most hunts of other monkey species result in only single captures.

Chimpanzees in general do not use tools to hunt, although they occasionally throw rocks or sticks at other animals.[13] Nevertheless, in at least one savanna population (Fongoli, Senegal), individuals of all ages and both sexes regularly use sticks to jab at prosimian quarry (bushbabies, *Galago senegalensis*) that nest in tree cavities.[14] Hunters ultimately capture their possibly immobilized victims by grabbing them with their hands. Like chimpanzees elsewhere, males in this community capture more prey in general than females do. However, females use tools to hunt more than males do, and the difference in percentage of bushbaby captures made by females versus males is greater than the percentage of prey captures made by females compared to males overall.[15] Tool use thus provides a wide range of individuals in this community with a means to acquire meat in a habitat where red colobus monkeys are absent, prey species diversity and density in general are low, and group hunting is less prevalent.[16]

Chimpanzees also occasionally steal or scavenge meat hunted by other predators. For example, the banana provisioning station at Gombe, which facilitated habituation and study in the early stages of this field project (Chapter 2), attracted both chimpanzees and baboons. Members of the two species were consequently in regular and often aggressive contact with each other.[17] During this era, chimpanzees were observed on twenty-seven occasions to steal freshly killed bushbuck fawns from predatory baboons.[18] Observers likewise saw chimpanzees at Taï steal three live red colobus monkeys from crowned hawk eagles that had not yet killed their victims.[19] Observations of chimpanzees scavenging previously killed and abandoned carcasses are rare, and most cases involve prey recently killed by themselves or other group members.[20]

SOCIAL AND ECOLOGICAL CONDITIONS THAT PROMOTE PREDATION ON RED COLOBUS MONKEYS

The frequency at which chimpanzees hunt red colobus monkeys, and the degree to which they are successful in making kills when they do hunt, vary widely both between different communities and within communities over time (Table 9.2). The sources of variation include differences in the rates at which chimpanzees and red colobus groups come into contact with one another; the size and composition of chimpanzee foraging parties that encounter these monkeys; food availability; and the size, composition, and location of prey groups. Analyses of these factors indicate that the relationship between a chimpanzee community and its preferred prey is highly dynamic and emphasize the need for long-term data to investigate the nature and consequences of chimpanzee predatory behavior.

Factors Affecting Encounter Rates

Encounter rates between chimpanzee and red colobus groups are fundamentally affected by the relative densities of the two species in a particular habitat. At Mahale, for example, increased red colobus numbers, resulting from habitat protection and improvement, appear to have been responsible for increased hunting frequencies by chimpanzees there over the course of several decades.[21] Conversely, drastically reduced red colobus numbers at Ngogo, apparently caused by chimpanzee predation, have resulted in reduced encounter rates between the two species and increased hunting of other primate species by the chimpanzees there.[22] It remains to be seen whether the predator–prey relationships between chimpanzees and red colobus at the various field sites are stable over time, since primate prey are not the principle source of nutrition for chimpanzees (discussed later).

Beyond baseline conditions of chimpanzee and red colobus population densities, the availability of plant food resources then influences whether foraging groups of the two species come into proximity with each other. Encounters resulting from food availability will largely depend on extrinsic factors affecting leaf and fruiting phenology in the forest. Chimpanzee community territories are large and typically extend over several red colobus group ranges. In their sometimes extensive travel in search of plant foods, therefore, chimpanzees may come into random contact with potential prey. In addition, although chimpanzee and red colobus diets are markedly different, the one dominated by ripe fruits and the other by leaves, there is some overlap in the plant species they utilize. Consequently, specific trees or groves of trees may attract groups from the two species simultaneously, bringing them into close contact and increasing hunting opportunities.[23]

Chimpanzee Male Party Size and the Likelihood of Hunting upon Encounter

Once they encounter a red colobus group, chimpanzees are more likely to hunt, and to make kills when they hunt, when they are in larger parties with more adult and adolescent males.[24] This is probably because with more males, who are the primary hunters, it is easier to disrupt the defenses of male red colobus, facilitating the capture of panicked, dispersing individuals.[25] Indeed, differences between communities and within communities over time in hunting and success rates are in some cases clearly due to differences in the number of males available to join in hunts.[26] However, larger parties with more males are in turn associated with both increased food availability and the presence of sexually receptive females. To clarify the proximate conditions that promote hunting, therefore, the respective effects of fruit abundance and sexual opportunity need to be disentangled.

Food Abundance and the Likelihood of Hunting upon Encounter

The most comprehensive analyses to date indicate that parties with multiple males are more likely to hunt when the ripe fruits they prefer are abundant in their home range.[27] These studies rely on estimates of fruit availability obtained by regularly monitoring the phenological status of hundreds of individual trees from species known to be preferred by chimpanzees in particular communities (Chapter 4, endnote 2). Earlier reports of monthly variation in hunting frequency identified an array of wet and rainy season predation peaks that differed both between and within field sites, and that consequently obscured the central importance of food abundance in promoting hunting behavior.[28] Since in tropical habitats the fruiting cycles of individual trees, as well as of tree species overall, are not reliably linked to rainfall alone, direct estimates of food availability have provided a more accurate means to evaluate the relationship between food abundance and predation.[29]

The positive association between ripe fruit availability and hunting propensity contradicts previous speculation that chimpanzees increase hunting rates to compensate for seasonal food shortages.[30]

Instead, it appears that, if anything, hunting is more likely when chimpanzees can readily meet their nutritional needs with plant-based foods.[31] Consideration of the potential costs associated with hunting helps explain this somewhat counterintuitive discovery. First, time spent searching for and then pursuing prey means less time to forage for plant foods, resulting in potential opportunity costs similar to those incurred during boundary patrols.[32] Second, and probably more importantly, capturing prey involves high-speed locomotion, climbing, and acrobatic arboreal pursuit, requiring an energy expenditure perhaps double that of ordinary running. Failed hunting attempts would obviously mean a net energy loss for all hunters, and even in successful group hunts many individuals might not obtain sufficient meat to offset their energetic expenditures (discussed later).[33]

The Presence of Swollen Females and the Likelihood of Hunting upon Encounter

As previously discussed, males are attracted to females with estrous swellings, and the presence of swollen females is consequently associated with larger party sizes (Chapter 7, endnote 3). Thus in principle swollen females could indirectly promote predation, insofar as their presence results in larger potential hunting parties.[34] However, males also face significant opportunity costs by hunting when swollen females are nearby, since they may miss chances to mate while they are engaged in time-consuming predatory behavior. Analyses of several long-term datasets indicate that, consistent with the prediction that males face a trade-off between hunting and mating, the probability that at least one individual will hunt in a party with a given number of males drops significantly if at least one swollen female is present.[35]

Red Colobus Group Characteristics and the Likelihood of Hunting upon Encounter

The location, size, and composition of red colobus groups also affect whether chimpanzees will initiate a hunt upon encounter. Early descriptions of hunting at Gombe suggested that chimpanzees were more successful at capturing red colobus when hunts took place in areas where the forest canopy was not continuous, for example where a comparatively tall tree extended above the surrounding tree crowns. This appears to be because when monkeys flee to such emergent trees, they are more easily trapped, since there are fewer additional escape routes available to them.[36] Further analyses at Gombe and Ngogo have subsequently demonstrated that chimpanzees at these sites are indeed more likely to initiate hunts both in areas of "broken" canopy as well as in lower, secondary growth forest.[37]

An accurate assessment of how the size and composition of a red colobus group influences the likelihood that a chimpanzee hunting party will attack it requires systematic observation of both predator and prey. This has only been attempted at Gombe. The results of this study indicated that chimpanzees are more likely to attack larger groups containing more immature individuals, a result consistent with the fact that immatures are their most frequent victims (Table 9.2). Larger groups also contain more adult males, increasing the number of potential defenders, but this does not necessarily prevent attacks.[38] Although similar studies monitoring both chimpanzee and red colobus groups have not been conducted at other field sites, a few published observations suggest that similar patterns may obtain elsewhere.[39]

THE QUESTION OF COOPERATION

Group hunts appear cooperative because more than one chimpanzee may be involved in the capture of a victim, and the probability that at least one monkey is captured increases when more hunters are present. However, it is not clear that hunters are acting cooperatively in the sense of intentionally helping each other to make a capture by coordinating their efforts in response to prey escape behaviors. Instead, it is likely that individual chimpanzees are essentially hunting independently from one another, each trying to capture a monkey for himself but incidentally gaining some advantage by doing so in the presence of other hunters.[40] For example, the presence of multiple hunters might simply promote a more chaotic flight response in a prey group, generally increasing their vulnerability. Lack of cooperation is also implied when successful hunters attempt to consume their prey alone (discussed later). An assessment of individual payoffs to hunters, as well as a determination of behavioral coordination between hunters, is therefore necessary to explore the degree to which hunts are cooperative.

Nutritional Benefits of Hunting in Groups

Estimates of the nutritional payoffs per individual in hunting groups of different sizes have been made for Gombe, Taï, and Ngogo, with mixed results. On the one hand, if the primary value of meat to chimpanzees is as a source of protein and/or fats, then cooperative hunting would be suggested if hunters on average obtained more meat when hunting in groups than when hunting alone.[41] The results from all three sites, however, indicate that larger hunting parties do not result in more meat per individual in the party.[42] This argues against cooperation at least by all individuals present, implying instead that some are "cheaters" benefiting from the efforts of their group mates.[43] On the other hand, if the primary value of meat to chimpanzees is as a source of important micronutrients rather than calories, cooperation would be suggested if the probability of acquiring at least some meat increased with increasing hunting party size. This relationship has been demonstrated at Gombe and Kanyawara, providing evidence of a nutritional payoff for cooperation.[44]

Is There Behavioral Coordination?

While quantitative estimates of prey meat availability to individuals in hunting parties provide some indirect support for cooperation, evidence of collaboration in pursuit behavior remains subjective and speculative. All observers report the impression that hunters are adept at evaluating the effects of other hunters on prey behavior and can effectively position themselves in the line of possible escape routes. However, since an individual hunter may be observed alternately chasing and intercepting prey, both on the ground and in the canopy, and in proximity to a variety of other hunters in rapid succession, it appears unlikely that individuals are systematically coordinating their efforts. More importantly, in view of the high pace and generally chaotic nature of group hunts, combined with the difficulty of documenting in detail behavioral events occurring high in the forest canopy and involving multiple predators and prey simultaneously, most researchers acknowledge the overwhelming difficulty of conducting empirical analyses to test subjective impressions of behavioral coordination.[45]

In addition, analyses of the triggering effect that especially avid hunters have on other males also cast doubt on the suggestion that social predation is complexly cooperative. At all five sites where hunting is frequent, observers have noted that some males appear more inclined to hunt than others, and in at least some communities the presence of specific males that frequently initiate hunts appears to increase the likelihood that group hunting will occur.[46] These so-called "impact hunters" seem to catalyze group hunts by being more willing to confront aggressive, counterattacking monkeys, causing prey groups to disperse and making individual monkeys more vulnerable to attack. When these males reduce their participation in hunting episodes or die, overall hunting and success rates within their communities decline.[47] This implies that successful group hunting relies on particular males disrupting prey group cohesion, thereby emboldening less eager hunters to join the hunting party, rather than on the coordination of individual pursuit behaviors.

SHARING THE KILL

The behavioral events that follow a successful hunt are extremely variable and depend on the identity of the individual that makes the kill, the size and composition of the hunting party, and the size of the carcass.[48] After group hunts, many individuals may attempt to acquire a portion of the kill. The meat possessor may be attacked and the entire carcass simply stolen – females and young males in particular appear especially vulnerable to outright theft.[49] Individuals competing for access to the meat possessor may interact aggressively with each other as well. Otherwise, group members cluster around meat possessors and engage in one or more of a wide variety of begging behaviors. These include sitting beside and staring intently at the possessor, reaching toward the possessor with the palm of the hand facing up, vocalizing, touching the possessor or the carcass, or actually pulling at the carcass (Plate 9). Possessors may then hand a portion of the kill to one or more beggars or passively allow a beggar to remove some. Occasionally meat possessors will also hand meat voluntarily to an individual that is not overtly begging.[50]

On the comparatively rare occasions that a lone hunter makes a kill, he or she is likely to consume it alone quietly or, in the case of mothers, share it with a dependent offspring. Similarly, after group hunts, meat possessors may try to escape quietly with carcasses from kill sites or retreat to branches or other locations where it is difficult for beggars to cluster around them. These observations suggest that possessors are disinclined to share.[51] Nevertheless, meat sharing is common, and several hypotheses have consequently been proposed to explain its functional significance. The initial premise is that individuals who share derive a benefit that outweighs the cost of surrendering their coveted and often hard-won food resource. Since sharing occurs primarily between unrelated individuals, kin-based benefits do not seem applicable.[52] Instead, three non-kin-based explanations have been examined empirically: sharing to reduce harassment while eating, sharing to form and maintain social bonds, and sharing to gain mating opportunities.

Sharing-under-Pressure[53]

As originally formulated based on early observations at Gombe, the "sharing-under-pressure hypothesis" was founded on the proposition that it is costly to defend meat from beggars in terms of time, energy, and potential injury. Meat possessors could therefore benefit if parting with a portion of a carcass meant that beggars would leave them in peace – in essence, "'paying' [them] to go away."[54] This implies that the relative benefit of sharing would depend on the size of the prey item. Since larger carcasses would include more meat than an individual could readily consume, surrendering bits would not represent a significant nutritional loss, and sharing would therefore result in a net benefit by reducing the costs associated with defensive behavior. By the same reasoning, small carcasses would not contain surplus meat, and thus sharing would be less beneficial since it would result in a greater nutritional forfeiture. The fact that smaller prey were often defended and consumed alone by possessors appeared consistent with this idea.

More recent support for the sharing-under-pressure hypothesis comes from a video analysis of meat sharing events at Gombe, in which the immediate effects of begging behavior on meat possessors, and the behavioral consequences of sharing for both possessors and beggars, were examined.[55] The results were as follows. First, the rate at which possessors consumed meat declined as the number of beggars increased. This indicated that begging was costly to possessors and suggested that it could be interpreted as "harassment." Second, the likelihood that a possessor would share increased with the duration and the intensity of the begging behavior – specifically, when beggars touched the carcass or possessor, or put their hand on the possessor's mouth, rather than just staring or reaching out. And finally, beggars were more likely to leave a begging cluster after they obtained meat than otherwise. Together these results suggest that sharing is an essentially selfish behavior that reduces the costs of harassment to meat possessors by increasing the probability that beggars will depart.

Sharing Meat to Help Form and Maintain Social Bonds

Since males are the most frequent meat possessors as well as the most frequent recipients of sharing, the social bonding hypothesis focuses on them in particular.[56] As previously described (Chapter 6), male chimpanzees form strong and lasting social bonds with each other, and meat sharing could serve to reinforce those bonds. If individuals incur a significant cost by sharing, which seems evident given the reluctance in general with which they relinquish meat, the behavior is expected to evolve only if they also receive (or have received) a benefit in return at some other time. This is in contrast to gaining the immediate payoff of reduced harassment, although the two functions are not necessarily mutually exclusive – even if harassment promotes sharing, it may not determine with whom an individual shares. Potential social or nutritional return benefits might include receiving grooming, support in conflicts, tolerance, or, indeed, meat.

There is quantitative evidence from three field sites supporting the idea that meat sharing facilitates male social bonding. Over a ten-year period at Mahale, the alpha male was observed to share meat more often with males that he groomed and associated with most.[57] An association between meat sharing and grooming among males more generally has since been demonstrated at Ngogo, where males tend to share meat more frequently with individuals from whom they receive more grooming, and likewise receive meat more often from males that they groom more often.[58] Males at Ngogo also share meat reciprocally, and they share more frequently with individuals with whom they frequently form coalitions, an association demonstrated at Taï as well.[59] By contrast, in the study of begging harassment at Gombe described previously, males did not cede meat more often to individuals with whom they groomed or associated more.[60]

As a final note, an intriguing study at Budongo has demonstrated that urinary oxytocin levels are higher in individuals after a variety of food-sharing events, including meat sharing, compared with after other feeding contexts or social grooming. Oxytocin is a hormone implicated in parent–offspring bonding and cooperative behavior among mammals, suggesting the possibility of an underlying neuro-biological basis for social bonding stimulated by meat sharing.[61]

Sharing Meat to Gain Matings

In an early study at Gombe, males were observed to share meat more often with swollen than with anestrous females.[62] This pattern of sharing later prompted the suggestion that by sharing with estrous females, males might receive additional matings as an immediate return benefit.[63] However, subsequent analyses of long-term data from Gombe, Kanyawara, and Ngogo have failed to demonstrate that males share preferentially with swollen females or to show that males gain extra matings when they do share meat with them.[64] A single study at Taï reported that females mated preferentially with males that had shared meat with them at least once over a twenty-two-month period.[65] However, this analysis neglected to consider the intensity of female begging behavior (endnote 52) or the influence of male aggression on female sexual behavior throughout the study period, the latter a critical shortcoming given the impact that male aggression has on females (see "Indirect Sexual Coercion" in Chapter 7).

IMPACT ON PREY POPULATIONS AND CONSERVATION IMPLICATIONS

Predation by chimpanzees can severely impact local monkey populations. Astonishingly, annual offtake rates for red colobus in some studies exceed 50 percent, although considerable variation exists between sites (Table 9.3). At Gombe between 1991 and 1993, red colobus groups were smaller toward the center of the chimpanzee community range compared with near the periphery.[66] Computer modeling at this time indicated that in the absence of chimpanzee hunting, the Gombe red colobus population would double in ten years.[67] At Ngogo, where hunting pressure is especially intense, the red colobus population within the chimpanzee community range declined by nearly 90 percent between 1975 and 2007, a decrease attributed to chimpanzee predation.[68] Consistent with this decline, encounter rates between Ngogo chimpanzees and red colobus groups also declined between 1998 and 2014.[69]

Significant reductions in red colobus population numbers point to the potentially important role chimpanzees may play in shaping the primate communities to which they belong. Computer modeling has indicated that chimpanzee predation on red colobus monkeys is not sustainable at either Gombe or Ngogo.[70] In contrast to cyclical top-down predator–prey systems, local red colobus populations appear vulnerable to extinction since they are not, apparently, a limiting resource

Table 9.3 *Estimates of the percentage of sympatric red colobus populations killed by chimpanzees annually at four long-term study sites*

Study site	Years	Percentage	Reference
Gombe	1973–1974	8–13%	Busse (1977: 908)
	1972–1975	20–41.6%	Wrangham and Bergmann Riss (1990: 165, 168)
	1982–1992	20–35%	Stanford (1995: 578)
Mahale	1983–1990	1.3–2.0%	Ihobe (2000, in Hosaka, 2015a: 277)
	1990–1995	3.6–5.8%	Ihobe (2000, in Hosaka, 2015a: 277)
Taï	1984–1995	3.2–7.6%	Boesch and Boesch-Achermann (2000: Table 8.5)
Ngogo	1995–1999	6–12%	Watts and Mitani (2002a)
	1995–2002	15–53%[1]	Teelen (2008)

[1] The much higher percentages compared to earlier Ngogo studies reflect a lower estimate of the red colobus population, made possible by systematic monitoring of four red colobus groups over a three-year period, combined with a higher hunting rate in later years (Teelen, 2008: 46).

for chimpanzees. Consistent with this possibility, several known red colobus groups have disappeared at Ngogo since the 1990s.[71] Local primate extinctions caused by human hunting are well documented.[72] The census and chimpanzee predation data from Gombe and Ngogo are the first evidence that a nonhuman primate species could lead to the local extermination of a sympatric primate group.

Nevertheless, if prey densities decline sufficiently and predators are able to switch to alternate prey, stable predator–prey cycles could result if prey populations can recover through reproduction or immigration. This appears possible at Ngogo, although by no means inevitable. On the one hand, hunting rates of red colobus there have declined in recent years as encounters toward the center of the community range have become less frequent and predation on other monkey species has increased.[73] Red colobus offtake rates, correspondingly, have declined. On the other hand, however, hunting of red colobus has increased in peripheral areas of the community territory, and neither the likelihood of hunting once encountering a group nor of killing at least one monkey in a hunt has declined. Thus predator pressure on red colobus monkeys at Ngogo remains considerable, even if reduced. Moreover, since males in neighboring chimpanzee communities are probably also hunting, the potential for population recovery through immigration into the Ngogo community range may be limited. These observations suggest that the long-term prospects for red colobus and alternate prey species at Ngogo remain uncertain.

The effectiveness with which groups of male chimpanzees can capture prey raises unexpected conservation questions. Can one endangered species cause the local extinction of another? What do we do if promoting the survival of one species negatively affects the prospects of another? How wide a net do we need to cast in order to safeguard a healthy forest ecosystem? Meanwhile, group hunting and meat sharing by chimpanzees are also intriguing from the perspective of human evolution, providing clues about ancestral predator–prey interactions, the evolution of the human dietary niche, and the roots of our unusual capacity for collective behavior and sharing among highly competitive individuals.[74] To gain a still richer understanding of the dynamic interplay of competition and cooperation that characterizes chimpanzee social existence, we need to examine the communicative behaviors that mediate their varied social interactions. This is the subject of Chapter 10.

10

Communication: The Form and Content of Signals

OVERVIEW

Chimpanzee social interactions involve a rich repertoire of body movements and postures, autonomic indicators of arousal, vocal and nonvocal sounds, and facial expressions. Scores of regularly observed expressive physical behaviors have been described for wild chimpanzees, and comprehensive catalogs continue to grow as new behaviors are identified, because they are either rare, site-specific, or recently invented. Similarly, building on a basic repertoire of calls described in early field studies, researchers continue to distinguish new vocalizations as acoustic analyses reveal subtle variants within previously delineated call categories. Nonvocal sounds, made by striking, scratching, or otherwise manipulating body parts or objects in the environment, are commonly produced and offer seemingly limitless opportunities for innovation.

Expressive behaviors can be categorized according to the sensory modality they involve, that is, either olfactory, tactile, visual, or auditory. They are considered communicative "signals" insofar as they appear to have evolved to convey information to other individuals and to mediate social interactions.[1] When signals are perceived through only a single sensory channel, they are comparatively easy to investigate. This is the case, for example, with loud vocalizations produced and monitored by individuals who are not in sight of one another. However, most chimpanzee interactions involve multimodal combinations of signals directed toward individuals that are in view. For example, it is common to observe a female approach and ultimately touch a male while she simultaneously crouches, bobs her head, and utters submissive vocalizations. Such multimodal events are more challenging to interpret, since it is difficult in practice to disentangle the functional consequences of their individual or variously combined components.

Animal communication is a social phenomenon, and analysis of communicative events necessarily requires consideration of both signalers and receivers. Classic descriptions of animal communication define the "message" of a signal as the information content of the signal itself. In this view, messages primarily consist of information about the sender. Potential information might include species, social group, individual identity, sex, status, physical condition, or motivational state. The "meaning" of a signal is its message combined with information available from the context of its production, and it is operationally defined in terms of the receiver's behavioral response. By combining message with context, receivers then have information available to predict the sender's probable imminent behavior and respond appropriately.[2] So, for example, high-ranking male chimpanzees utter "pant hoot" vocalizations both as they approach a food tree and after they have been calmly feeding in it for some time. A female already feeding in the tree might respond to the former by uttering submissive "pant grunts," since males often charge individuals when they arrive at a feeding location, but ignore the latter.

An understanding of communicative signals provides a basis for describing how individual social relationships are established and maintained, as well as for exploring how social organization more generally is patterned and perpetuated. A considerable amount of research has also focused on whether chimpanzee communicative behavior exhibits structural, interactive, or cognitive characteristics that resemble aspects of human language and might therefore offer insights into the nature of the linguistic precursors that would have been present in human ancestors. Areas of interest include whether receivers may gain information from signals about conditions or events external to the sender, how much control individuals have over signal production, the degree to which signal production is affected by the identity of receivers and may be "intentional," and whether signal production and use are influenced by social learning.

OLFACTORY COMMUNICATION

As in other catarrhine primates, olfactory communication appears comparatively unimportant in chimpanzees (Chapter 1). The most commonly observed olfactory behavior of a social nature occurs during "sexual inspections," when a male will closely peer at and bring his nose close to a female's genital area. The male may also touch the female's vulva with his finger and then smell his finger. These behaviors are also occasionally exhibited by females and presumably represent efforts to monitor chemical cues informative of a female's reproductive condition. In addition, individuals of either sex may touch the genitals of a male and then smell their fingers, and they are sometimes observed sniffing the ground when searching, apparently, for other group members. It is not known what specific information individuals acquire in such cases.[3]

TACTILE COMMUNICATION

Touch plays a central role in chimpanzee social life. As previously noted, social grooming – gently picking through the hair on the head, torso, or limbs of another individual – appears to be essential to forming and maintaining affiliative social relationships. Mothers maintain uninterrupted contact with their infants for several months after birth and continue to groom them regularly and provide comforting physical contact for many years after.[4] Older individuals may spend hours daily grooming and being groomed. A variety of briefer contact behaviors – touching, patting, kissing, and embracing – provide reassurance in stressful contexts, for example when hearing vocalizations from neighbors or while witnessing within-group conflicts. Likewise, friendly physical contact after aggressive confrontations calms fight participants and promotes tolerance between them, while contact between antagonists and a third party may comfort fight participants as well as bystanders (Chapter 1).

Physical contact can obviously also be aggressive and serve to inflict pain. A comparatively gentle slap or push may simply cause the recipient to move away. For example, an adult resting on the ground might push away a playful youngster that is jostling against him. In such cases, the distinction between threat and play, which itself involves a wide range of physical contacts, may be ambiguous. Failing to heed the signal, the juvenile may continue to make inadvertent contact with the adult until a less ambiguous but nonetheless fairly mild response is elicited. Alternatively, aggressive contact can be severe and prolonged, involving slapping, kicking, biting, twisting limbs, and dragging on the ground. Maturing males gradually establish dominance over all community females through physical aggression, and among adults of both sexes dominance relationships are likewise established and maintained through aggressive physical contact that at times is severe.

VISUAL COMMUNICATION

Signals that are perceived visually fall along a continuum ranging from some that are completely involuntary to others that appear intentional or goal-directed, in the sense of one individual intending to produce a specific effect in another. Commonly observed involuntary visual signals include female sexual swellings and piloerection (also called "bristling" at Gombe[5]). Many facial expressions seem tightly linked to emotional state and largely involuntary as well, although individuals may be aware of their signal value despite being unable to control them. At the other end of the spectrum, a wide range of limb and head movements produced during social interactions exhibit features characteristic of first-order intentional communication.[6]

Involuntary Signals

The size of a female's sexual swelling provides reasonably precise information about the timing of ovulation and is closely correlated with male mating effort (Chapter 7, endnote 9). As with exaggerated sexual swellings in other primate species, the adaptive value of this signal to female chimpanzees is unclear. Swelling size could be a reliable indicator of female quality and attract the best males;[7] swellings could permit males to track female fertility across sexual cycles;[8] or swellings could be graded signals that attract the highest-ranking males when conception is most likely, but also encourage mating by lower-ranking males when conception is possible but less probable.[9] The benefit in the latter case would be in causing paternity uncertainty and thereby selecting against the evolution of infanticidal behavior in males, while also increasing the probability of being fertilized by a high-quality individual (Chapter 8).

Piloerection, or erection of the body hair, is common in highly aroused individuals, especially during aggressive interactions. Receivers presumably learn that piloerection is indicative of elevated arousal through its association with aggressive behavior and respond to it accordingly. For example, charging displays by males are often preceded by rocking back and forth with hair on end, so piloerection in this context might put nearby chimpanzees on their guard. Likewise, piloerection in a high-ranking male sitting beside an estrous female might elicit submissive behavior in a lower-ranking male walking nearby. By contrast, piloerection in an adult male on a boundary patrol, subsequent to hearing a neighbor's calls, might help arouse a nearby ally and elicit a supportive reassurance embrace from that individual.

Chimpanzees have a number of stereotypical facial expressions that are consistently associated with particular contexts and, presumably, emotional states (Table 10.1). Some are linked to specific vocalizations and are likely a direct consequence of adjusting the contours of the lips and mouth to produce different sounds (Fig. 10.1, Plate 10).[10] The signal value of these facial configurations is probably subordinate to that of the vocalization itself in most cases. Others, such as the "full closed grin," are inferred to be closely linked to the emotional state of individuals and appear to be completely or largely involuntary. Few systematic studies have been devoted to identifying how specific facial expressions affect receivers. Meanwhile, several studies of captive chimpanzees have explored whether some of their facial expressions might be homologous with structurally similar ones exhibited by humans.[11]

It is important to note that while an expressive behavior may be involuntary, individuals may nevertheless be aware of its signal value to receivers. Observations of individuals in captivity apparently attempting to hide expressive behaviors from other group members support both inferences, that is, that individuals are unable to control some communicative signals and that they are also aware of the potential effect of these signals on receivers. For example, status contests between adult males can involve multiple aggressive as well as conciliatory interactions transpiring over periods of months. In a well-documented dominance struggle between three males in Burgers' Zoo (Arnhem, the Netherlands), one male was observed to cover his "full closed grin" with his hands while turning away from his opponent in the midst of a charging display. On a different occasion, a lower-ranking adult male in the vicinity of an estrous female was observed covering his erect penis with his hands as a higher-ranking male passed near.[12]

Intentional Gestures

Intentionality is inferred when signals are directed at specific receivers, signal production depends on the attention state of the receiver, and the sender persists in signaling if, after a brief pause to monitor the receiver, an adequate response is not received.[13] Wild chimpanzees have many head, limb, and body movements that meet these criteria

Table 10.1 *Chimpanzee facial expressions and associated contexts in which they are commonly observed*

Goodall (1986)[1]	Nishida et al. (2010)[2]	Common context	Common accompanying call
Relaxed face	Relaxed face	In calm social or resting contexts	
Relaxed face drooped lip	<No entry>	Idiosyncratic variation of relaxed face	
Lip flip	Lip flip	Unspecified	
Sneer	Sneer	When suddenly alarmed, especially around humans	
Pout face	Pout	Infants seeking nipple, slipping; older individual begging (Goodall, 1968b)	Hoo
Horizontal pout	<— includes	When juveniles lose mothers; after threat or attack	Whimper
Grinning[3]	Grin	After threat or attack; by females copulating	Squeaks, screams
Low open grin[4]	Grin-low-open		
Full open grin	Grin-full-open		
Low closed grin	Grin-low-closed		
Full closed grin[5]	Grin-full-closed	Frightening stimulus, e.g., seeing or hearing neighbors; following threats, perhaps signaling benign intent	No call
Compressed-lips face[6]	Compress lips	When "glaring" at someone, sometimes prior to attack or copulation	

Play face[7]	Play face	With nonaggressive physical contact	Sometimes laughter
Hoot face[8]	Hoot face	Lips pushed forward	With pant hoot variants
<No entry>	Funny face	Idiosyncratic, possibly a reassurance signal	
<No entry>	Protrude tongue	Infants, significance unclear	
<No entry>	Pucker cheek	Suck in cheeks, significance unclear	

[1] Terminology following Goodall (1986: Fig. 6.1, photos pp. 121–123, expanded from Goodall, 1968b: Table 12–2), based on observation of Gombe chimpanzees. Goodall (1968b: 323) notes that there are expressions intermediate between all the main types. Plooij (1980: Appendix A) uses identical or closely similar terms. Similar forms of many of these facial expressions occur in monkeys and other apes, e.g., the relaxed face, the pout face, grinning, and the play face (van Hooff, 1967).

[2] The Nishida et al. (2010) ethogram mainly follows Goodall's (1986) terminology. Where it differs, equivalent terms are listed.

[3] For grinning, "full/low" refers to amount of teeth showing, and "open/closed" refers to whether the mouth is open.

[4] Goodall (1989), cited in Nishida et al. (2010).

[5] Goodall (1968b) used the term "silent grin" when grinning occurred without a vocalization. This is synonymous with van Hooff's (1967) term "silent bared-teeth face." Waller and Dunbar (2005: Table 1) also list "fear grin" and "fear grimace" as synonyms.

[6] Probably the same as the general primate "tense-mouth face" (van Hooff, 1967).

[7] van Hooff (1967) used the term "relaxed open-mouth face" for play face.

[8] Goodall (1968b: Table 12–2).

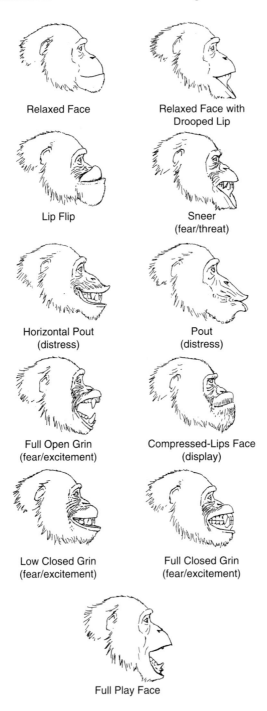

Relaxed Face

Relaxed Face with
Drooped Lip

Lip Flip

Sneer
(fear/threat)

Horizontal Pout
(distress)

Pout
(distress)

Full Open Grin
(fear/excitement)

Compressed-Lips Face
(display)

Low Closed Grin
(fear/excitement)

Full Closed Grin
(fear/excitement)

Full Play Face

Fig. 10.1 Facial expressions of chimpanzees. Reprinted from
The Chimpanzees of Gombe: Patterns of Behavior (Goodall, 1986: Fig. 6.1), with
permission.

(Table 10.2). Consistent with reports from captive studies, gesture type varies depending on whether the receiver is attending to the sender: gestures involving sound or contact are more common when receivers are not attending, whereas silent gestures are more common when individuals are in visual contact.[14] For example, an individual begging for food will walk around in front of the food possessor before reaching out a hand, palm up. Similarly, in a comparative study of mother–infant pairs from Kanyawara and Taï, individuals engaged in turn-taking gestural sequences in which gaze and body orientation functioned to establish a "participation structure," after which a range of contact and movement behaviors were exchanged to initiate joint travel.[15]

Anecdotal accounts of chimpanzees apparently deceiving other individuals provide additional evidence of intentionality in gesture use. For example, banana provisioning at Gombe generated intense feeding competition, such that lower-ranking individuals were sometimes unable to acquire fruits when dominant individuals were present. One young male was observed on several occasions to succeed in initiating group travel away from the feeding station by walking energetically out of camp, only to then circle quietly around and return to obtain a share of bananas.[16] During the course of the dominance struggle at Burgers' Zoo mentioned previously, one male was observed to mildly injure his hand. For the following week he walked with a limp whenever he was in view of his opponent, but otherwise walked normally. The researchers interpreted this behavior as having been intended to temporarily reduce the likelihood of further fighting.[17]

The degree to which gesture production is influenced by social learning is unclear. Gesture use is flexible, in that a given gesture might be deployed in many contexts, many different gestures may be produced in a single general context, and individuals vary in their use of gestures. Captive chimpanzees in particular also occasionally produce idiosyncratic gestures. Taken together, this flexibility has been adduced as evidence that gesture production and use are to a large extent learned by individuals through a process of "ontogenic ritualization" over the course of repeated interactions with other group members.[18] Nevertheless, observations from long-term studies in the wild reveal a maximum repertoire of sixty to seventy types that appear largely consistent across field sites, with an absence of idiosyncratic forms reported in at least one population.[19] Additionally, a study at Budongo demonstrated that compared with young individuals, adults were more often able to elicit responses from receivers with single gestures and less often resorted to persistent signaling with multiple

Table 10.2 *Some common manual and postural gestures of wild chimpanzees in various behavioral contexts*[1]

Expressive movement	Description[2]
Threatening	*Gestures that elicit submissive behavior in the individual they are directed toward*
Glaring (stare fixedly)	Lips compressed, animal stares fixedly at another individual
Head tipping (tip head)	Slight backward jerk of the head, invariably accompanied by the "soft bark"
Arm raising (raise arm quickly)	Either forearm or entire arm raised rapidly, palm toward threatened animal
Hitting away (hit toward)	Hitting movement with the back of the hand directed toward threatened animal
Flapping (flap)	Downward slapping movement of the hand in direction of threatened animal
Branching (shake branch)	Taking hold of branch and shaking side to side or backward/forward
Stamping and slapping (stamp/slap)	Part of charging display, stamp feet and slap hands on ground
Bipedal arm waving and running (run bipedal)	A form of charging display
Bipedal swagger (hunch bipedal)	Stand upright, arms out to side, swaying foot to foot, shoulders hunched
Sitting hunch (hunch sit)	While sitting, hunch shoulders and raise arms in front or to the side
Rocking (rock side to side)[3]	Rhythmic side-to-side movement of torso when sitting, in preparation for a charging display
Quadrudedal hunch (hunch quadrupedal)	Standing on all fours, back rounded, head pulled between shoulders, sometimes followed by attack

Charging display[4]	Galloping with a variety of threatening movements
Attacking charge	Typically a silent, high-speed charge followed by physical attack
Submissive	*Behaviors by a subordinate individual directed at a higher-ranking individual when either (i) the dominant has threatened or attacked the subordinate, (ii) the dominant has merely passed near or approached the subordinate, or (iii) the subordinate has approached or passed near the dominant*
Presenting (present with limbs extended)	In quadrupedal stance, turning rump toward higher-ranking individual
Bowing, bobbing, crouching	A variety of up-and-down body movements while facing a dominant in which the subordinate keeps the body low and roughly horizontal while flexing the limbs to various degrees, typically accompanied by submissive vocalizations
Touch	Reaching out and touching with the hand
Reach without touching (extend hand)	Reaching out without touching, usually with palm up
Kiss	Pressing lips or teeth to body of another, including mouth-to-mouth contact
Mount	Embrace from behind, chest touching back, usually with pelvic thrusting
Wrist bending (wrist toward)	With wrist flexed, holding out back of hand or wrist toward lips of dominant
Bending away	Leaning away while flexing elbow and wrist and drawing arm to body

Table 10.2 (cont.)

Expressive movement	Description
Greeting	*Primarily submissive and reassurance behaviors typically observed between individuals that are reuniting after a period of separation (i.e., during "reunions"; see Chapter 6)*
Reassuring a subordinate	*Behaviors by a dominant individual toward a subordinate in response to a submissive gesture directed toward him or her*
Touch	In response to presenting or bowing/crouching
Pat	Patting with the hand or fingers, palm down, in response to extreme agitation
Embrace	Embracing with one or both arms from the front, side, or back of an individual
Mount	As in submission, in response to screaming or presenting
Kiss	As in submission, in response to reaching or crouching
Rump turning (present)	Like presenting, may be followed by the two individuals pressing their rumps together
Reassuring self in contexts of social excitement or fear	
Touch, embrace, etc.	With nearby individuals of any rank
Movements suggesting that the individual is experiencing stress, anxiety, or frustration	
Scratching (scratch self)	Rake nails across skin of arm or torso, usually with increasing vigor in conflict situations

Yawning	Yawn repeatedly, either toward or away from a mild disturbance
Masturbation	Hold or toy with erect penis
Temper tantrum (throw temper tantrum)	By infants and juveniles: screaming, leap in air, hurl self on ground, writhe, hit self against objects
Redirected aggression	Threaten or attack a subordinate individual after being attacked or threatened by a dominant individual
Soliciting support	*When individuals are in conflict situations they may direct gestures toward allies, sometimes while screaming, looking back and forth between ally and rival.*
Extend hand[5]	Extend arm horizontally, wrist and fingers extended, palm up or down
Embrace, touch, mount	Similar to gestures described for reassurance
In grooming context	*Gestures that appear to elicit grooming by another individual*
Present for grooming (solicit grooming)	Stand or sit by, head slightly bowed
(Raise arm slowly)[6]	Extend arm overhead
In play context	*Youngsters initiate social play with a variety of contact behaviors (hit and run, slap, tickling, pulling fingers or toes, etc.). Several postures and expressions are common once individuals are engaged in play.*
Play face	Lower lip retracted exposing lower teeth; mouth open and increased retraction with more vigorous play

Table 10.2 (cont.)

Expressive movement	Description
Play walk	Short, stilted steps with rounded back, head bent down and pulled back between shoulders
Food sharing	*Gestures that may result in food sharing*
Reach out hand (extend hand to beg)	Palm up, toward a food possessor, either toward their hand or mouth
Touch lips or food	As earlier in table, but actually touching the possessor or the food
Male courtship displays (solicit copulation)	*Attention-getting behaviors observed prior to mating and accompanied by penile erection[7]*
Male invite (open thighs)[8]	Sit with thighs apart, looking toward female, sometimes flicking penis
Beckoning	Sweeping movement with arm while standing bipedally
Leaf clipping (clip leaf)[9]	Pull leaf between teeth or lips, or tear with hand, making conspicuous sound
Glaring (stare fixedly)	Described earlier in table, presumably with a threatening component in sexual context
Bipedal swagger (hunch bipedal)	Described earlier in table, presumably with a threatening component in sexual context
Sitting hunch (hunch sit)	Described earlier in table, presumably with a threatening component in sexual context
Branching (shake branch)	Described earlier in table, presumably with a threatening component in sexual context
Female courtship displays (solicit copulation)	*Behaviors indicating readiness to copulate*

Presenting (present with limbs flexed)

Approach, turn, and flatten body to ground with hindquarters toward male, look back over shoulder

Other behaviors in sexual context

Inspection (inspect genitals)

Male peers closely at, smells, or touches female vaginal area and smells fingers, usually when the female is not fully swollen

---

1. The organization and content of this table follow Goodall (1968b, 1986). Movement terms in parentheses are the ones used in Nishida *et al.*'s (2010) updated and greatly expanded online ethogram, which includes video clip examples. Note that most gestures are observed in a wider variety of contexts than listed and with much individual variation in their expression. Reynolds and Reynolds (1965) at Budongo and Nishida (1970) at Mahale provided similar early descriptions of many of these common expressive movements. Hobaiter and Byrne (2011a) and Roberts *et al.* (2014) describe recently compiled repertoires from Budongo. See Call and Tomasello (2007) for a catalog of captive chimpanzee gestures.

2. Descriptions are from Goodall (1968b) unless otherwise noted.

3. In Goodall (1989, cited in Nishida *et al.*, 2010: 158).

4. Charging displays potentially involve a variety of components, including branch shaking, stamping, slapping, bipedal swaggering, dragging branches, throwing objects, and pant hoot vocalizations (see Goodall, 1986: 316, for a detailed description). Hosaka (2015b) refers to behaviors involving any combination of these components as "intimidation displays."

5. In Goodall (1989, cited in Nishida *et al.*, 2010: 73).

6. In Nishida *et al.* (2010: 149).

7. A chimpanzee's erect penis is bright red and highly visible against his pale thighs.

8. Goodall (1986: 447).

9. Absent at Gombe but observed at Mahale, Taï N, and Kanyawara (Whiten *et al.*, 1999).

types. These results suggest that chimpanzees have a finite repertoire of species-typical gesture types, but that with experience individuals improve their ability to deploy them, a possibility dubbed "repertoire tuning."[20]

AUDITORY COMMUNICATION 1: VOCAL SIGNALS

Chimpanzees produce a wide range of vocal and nonvocal auditory signals. Most systematic research has focused on vocal behavior, both because of its importance for understanding the role of vocalizations in mediating chimpanzee social relationships and because of its potential relevance for theorizing about the evolution of language. Nonvocal acoustic signals include sounds chimpanzees generate with the mouth, by scratching themselves, or by shaking or striking objects. Among nonvocal auditory signals, only drumming on tree buttresses has been the subject of multiple quantitative studies at several field sites. As with gestures, most auditory signals are produced in a variety of contexts, and many different signals can be produced in any given context. Auditory signals appear primarily to provide information about senders themselves, although in conjunction with contextual information receivers may infer additional information as well.

The chimpanzee vocal repertoire is estimated to include between thirteen and thirty or so call types (Table 10.3). Uncertainty about the number of call types reflects the difficulty of delineating categories in the face of tremendous variability, both within and between individuals, in the acoustics of calls. In addition, as is the case in other primate species, many chimpanzee vocalizations "grade" into one another. This means that acoustic features vary in a continuous fashion between some types, such that it is possible to "connect" the different types with intermediate forms; indeed, in some cases intermediate forms between "types" are produced more often than the designated "types" themselves.[21] Whether different call variants are functionally significant to the chimpanzees is an area of ongoing research (for example, different "pant hoot," "bark," or "grunt" variants).

Although groups of chimpanzees sometimes produce piercingly loud, varied, and prolonged bouts of calls and call choruses, seemingly dominating the surrounding acoustic environment, systematic focal animal sampling has revealed that individuals in fact vocalize at very low rates overall. A female foraging alone with an infant may remain silent or nearly so for an entire day. Females and other low-ranking individuals foraging and traveling in small parties might utter only the

Table 10.3 *The wild chimpanzee vocal repertoire documented from behavioral observation and acoustic analysis*

Goodall (1986)[1]	Nishida et al. (2010)[2]	Marler (1976)[3]	Additional variants[4]	Descriptions and contexts excerpted from Goodall (1986)[5]
<Pant hoot> (ph)	1 Pant-hoot	1 Pant hoot		Highly variable long-distance calls with distinct phases, given in a wide range of contexts, primarily by adult males
1 Arrival ph	<— includes	<— includes	Food arrival ph[6]	Travel and arrival at large food trees, reunions
2 Inquiring ph	<— includes	<— includes		Travel, exchanges with individuals out of view
3 Spontaneous ph	<— includes	<— includes		Peaceful feeding and resting
4 Roar ph	2 Hoot	<— includes		Charging displays, neighbor contact
5 Bark	3 Bark	2 Bark		Loud, variable, sharp calls during social excitement
Includes —>	Includes —>	Includes —>	Hunting bark[7]	When hunting, but also other contexts
Includes —>	Includes —>	Includes —>	Snake bark[7]	When seeing a snake, but also other contexts
6 Cough-threat	4 Cough bark	3 Cough		Soft bark, uttered as mild threat at subordinate
7 Waa bark	5 Waa bark	4 Waa bark		Loud, sharp calls uttered in agonistic contexts
8 Pant grunt	6 Pant grunt	5 Pant grunt		By subordinate approaching or being approached by dominant
9 Pant bark	7 Pant bark	<— includes		With higher arousal or fear, pant grunts may grade into pant barks

Table 10.3 (cont.)

Goodall (1986)	Nishida et al. (2010)	Marler (1976)	Additional variants	Descriptions and contexts excerpted from Goodall (1986)
10 Pant scream	<– includes	<– includes		At still higher intensity, pant barks may grade into pant screams
11 Whimper	8 Whimper	6 Whimper		Soft, lower-pitched sounds given in distress by subordinate individuals, may progress into squeaks and/or screams
12 Squeak		7 Squeak		Short, high-pitched, delivered in sequences in response to threats from a dominant individual; may grade into screams
13 Crying	9 Whimper-scream			A combination of whimpering and screaming in distress
<Scream>	10 Scream	8 Scream		Highly variable, high-pitched, acoustically noisy calls delivered in long sequences during extreme distress, for example after being attacked
14 Victim scream	<– includes	<– includes		Harsh and prolonged, when a chimpanzee is being attacked
15 SOS scream	<– includes	<– includes		After attack; clear, high-pitched, to solicit support
16 Tantrum scream	<– includes[8]	<– includes		Infants during weaning, sometimes adults after an attack; loud, harsh, and sometimes resulting in gagging

Call		Call		Specific call	Context
17 Hoo	Includes →	11 Hoo	(with whimper)	Aggressor scream[9]	By lower-ranking individuals, while attacking or threatening another individual
	Includes →		Includes →	Resting hoo[10]	Initiating travel, resting, infants reestablishing contact with mothers (can grade into whimpering)
	Includes →		Includes →	Travel hoo[10]	
18 Huu[11]	Includes →	12 Huu	(with whimper)		Puzzlement, surprise, slight anxiety toward small snakes, unidentified rustlings, dead animals, etc.
	Includes →		Includes →	Soft huu[12]	Newborn's response to sudden movement by mother
	Includes →		Includes →	Alarm huu[12]	
		13 Staccato call[13]			
19 Food grunt[14]	9 Rough grunt	14 Grunt, food			Soft, while feeding, can be uttered in exceptionally long sequences including >100 calls
20 Food aaa	<— includes	15 Grunt, aha			While feeding, higher-pitched and loud, also in very long sequences, often interspersed with food grunts

Table 10.3 (*cont.*)

Goodall (1986)	Nishida et al. (2010)	Marler (1976)	Additional variants	Descriptions and contexts excerpted from Goodall (1986)
<Grunt>	16 Grunt	10 Grunt		Variable, low-pitched, acoustically noisy; range of social contexts
21 Soft grunt	<-- includes	<-- includes		During foraging and travel among friendly individuals
22 Extended grunt	17 Grunt, extended	<-- includes		Long, draw-out grunt while resting
23 Nest grunt	18 Grunt in bed	<-- includes		Soft double grunt when looking for a nesting site
24 Pant	19 Pant	11 Pant		Breathy, rapid inhalation and exhalation, in greeting, grooming
25 Copulation pant	20 Pant in copulation	<-- includes		By males when mating, similar to laughter, but more rapid
26 Copulation scream	21 Squeal in copulation	(with scream?)		By females when mating, high-pitched, squeak-like
27 Laughter	22 Play pant	12 Laughter		Soft, breathy pants during play, tickling
Includes -->	Includes -->	Includes -->	Spontaneous[15]	
Includes -->	Includes -->	Includes -->	Laugh-elicited[15]	
28 Wraaa	23 Wraaa	13 Wraaa		Loud, drawn-out alarm call for snakes, buffalo, predators; also to abnormal behavior in a group member, e.g., sickness or death

1 Call types from Goodall (1986), based on behavioral observation, following original descriptions in Goodall (1968b: Table 12–2). The original term "bobbing pants" was changed to "pant grunts" by Marler (1976), which was subsequently adopted by Goodall (1986). Female copulation screams were called "squeaks" by Goodall (1968b) and Nishida (1997); Tutin and McGrew (1973: Table III) called them "squeaks." Reynolds and Reynolds (1965: Table 11–7) also provided early descriptions of the vocal repertoire along with audio spectrograms.

2 The Nishida et al. (2010) ethogram mainly follows Goodall's (1986) terminology, combining some closely similar call variants.

3 Based on acoustic analyses of tape recordings made at the Gombe feeding station, Marler (1976) combined many of Goodall's (1968b) original call types, on the grounds that they either were indistinct acoustically or graded into each other. This reduced the vocal repertoire to thirteen call types. Marler and Tenaza (1977) provide acoustic details and spectrograms of the thirteen call types.

4 Several studies subsequent to Marler (1976) and Marler and Tenaza's (1977) original acoustic analyses have documented acoustic differences between some closely similar calls and related them to contextual variation in call production.

5 Brief descriptions of typical call contexts are from Goodall (1986: 129–131, 134–136) and refer to the calls she distinguished by observation (column 1).

6 Originally called "food-pant-hoots" by Wrangham (1975: 3.34). Clark (Arcadi) and Wrangham (1993) at Kanyawara and Notman and Rendall (2005) at Budongo identified distinctive acoustic characteristics of pant hoots produced upon arrival at food trees. Notman and Rendall (2005) also identified distinctive acoustic features of pant hoots produced by chimpanzees traveling on the ground. An apparently rare, idiosyncratic pant hoot, called a "whimper-hoot," has been documented at Gombe (Marler and Hobbett, 1975: 103) and Kanyawara (Arcadi, 1996: Fig. 5).

7 Crockford and Boesch (2003), at Taï NM, identified two additional bark types based on acoustic analysis. Barks produced in the contexts of hunting and upon seeing snakes differed from each other as well as from barks produced during aggression, in response to calls from community members out of sight, in response to hearing neighbors, and when traveling. Barks produced in the latter contexts were not acoustically distinct from each other.

8 Nishida et al. (2010) also list "choke in tantrum" (following Goodall, 1986: 130), referring to presumed "glottal cramps" when, during "loud and prolonged tantrum screaming, the chimpanzee may seem to choke and only hoarse squeaks or rasping sounds can be heard."

9 Slocombe and Zuberbühler (2005a), at Budongo, identified an acoustically distinct scream produced primarily by low-ranking males, females, or juveniles while they were "chasing, beating, or threatening an individual" (p. 71).

10 Based on a sample of twenty calls from two Budongo individuals, Gruber and Zuberbühler (2013) found that hoos produced before travel were shorter, lower-pitched, and delivered in longer sequences compared to hoos produced while resting.

11 Called "alert hoo" by Crockford *et al.* (2012).

12 Schel *et al.* (2014) distinguished two "huu" vocalizations at Budongo, a "soft huu" produced whether or not other individuals were present, and an "alarm huu" (acoustically longer, louder, and higher-pitched than the "soft huu") possibly produced preferentially in the presence of other individuals. However, their sample of proposed "alarm huu" calls was not large enough to statistically confirm this contextual distinction (p. 7).

13 Nishida *et al.* (2010), following Plooij (1980).

14 Food grunts are short, variable vocalizations often uttered in long sequences, sometimes in excess of 100 calls, rising and falling in pitch (Arcadi, unpublished data). They were simply called "grunts" by Goodall (1968b: 30). The vocalization was renamed "rough grunts" by Marler (1976) and then "food grunts" by Goodall (1986).

15 Davila-Ross *et al.* (2011) identified two acoustically distinct laughter types among chimpanzees living in the Chimfunshi chimpanzee sanctuary in Zambia. "Spontaneous laughter" occurred in the absence of a playmate's laughter, whereas "laugh-elicited laughter" followed the laughter of a playmate within five seconds. Play bouts were significantly longer when a play partner produced spontaneous laughter compared with silent play, and play bouts were significantly longer when both partners produced laughter compared with bouts in which only spontaneous laughter occurred.

occasional contact or submissive vocalization over the course of a day. Adult males utter more non-submissive calls than other classes of individuals, but they are nonetheless frequently quiet for extended periods as well, and overall they only produce two to four vocalizations per hour on average. Adult males utter pant hoots, loud calls exchanged between individuals out of view and their most frequent vocalization, at rates of less than 1.0 to 2.5 calls per hour.[22]

Information Content of Vocalizations 1: Sender Characteristics

Individual Identity

Pant hoots are loud calls primarily produced by adult males, frequently before, during, and immediately subsequent to travel. They thus appear important for maintaining contact and regulating spacing between individuals out of view.[23] Individual distinctiveness in these calls would provide useful information to listeners if it improved their ability to choose with whom to associate at a given time. Numerous acoustic analyses have, after controlling for within-individual variation, indeed revealed consistent differences between the pant hoots of different males.[24] Long-time researchers also report the ability to identify particular individuals by their pant hoots. Nevertheless, no empirical study in the wild has demonstrated that individuals recognize each other by their calls, although an experimental study in captivity indicated that one test subject was able to recognize the pant hoots (as well as "pant grunts" and "screams") of her captive group mates.[25]

Isolated studies have pointed to vocal signatures in a few other chimpanzee calls that are audible to distant listeners. In an experimental study at Budongo, researchers played recordings of aggressive "barks" to individuals that had previously been involved in unreconciled aggressive conflicts. Listeners were more likely to move away from the hidden playback speaker if the call was from an individual that had a close social bond to the listener's original opponent, suggesting the ability to identify the caller's identity. Research at Kanyanchu, a site in Kibale National Park developed for tourism, indicated that both the "whimpers" and "screams" of three-year-olds were individually distinctive, potentially making it possible for mothers to locate their offspring when they are out of view and in distress.[26] Finally, an acoustic analysis indicated that female "copulation screams" at

Budongo are individually distinctive, although it is unknown whether listeners actually identify females based on these calls.[27]

Age and Sex

Vocal development in chimpanzees has received comparatively little systematic study. Nevertheless, distinctive "staccato" and "grunt" calls produced by infants have been described both in the wild and in captivity and would provide cues of age.[28] There is evidence for sex differences in two call types produced by adults, pant hoots and screams. Pant hoots typically consist of four parts, each containing one or more elements: (1) a quiet, tonal, "introduction" phase; (2) a "build-up" phase increasing in loudness and voiced on both inhalation and exhalation; (3) a "climax" phase composed of higher-pitched, scream-like calls; and (4) a "let-down" phase, resembling the build-up but decreasing in intensity. A study at Gombe found that female pant hoots were longer in duration and lacked the climax phase present in male calls.[29] Screams are noisy, high-pitched calls that occur in a range of conflict contexts. A study at Mahale showed that the screams of females are higher in pitch than those of males.[30]

Status and Physical Condition

Call usage varies between individual chimpanzees, and some of this variation reflects patterns of dominance within communities. For example, pant grunts are directed by lower-ranking individuals to higher-ranking ones, typically upon reunion, and constitute a reliable indicator of relative dominance status. Since pant grunts are also sometimes produced during travel, grooming, feeding, and resting, contexts in which the relative status of the sender and receiver appear to be evident, it is unclear whether producing the call functions to affirm dominance status or to test the receiver's tolerance for proximity.[31] In contrast, while as mentioned previously pant hoots are the most frequent call produced by adult males generally, high-ranking males produce them at higher rates than individuals of all other age/sex classes.[32] To the extent that this call promotes the reunion of separated individuals, the disproportionate production of pant hoots by high-ranking individuals may reflect their superior ability to tolerate feeding or mate competition, thereby potentially providing information about status.[33]

Several studies from Kanyawara have explored the possibility that the comparative ability to produce clear, high-pitched pant hoot climax elements provides a cue about the physical condition of the caller and, by extension, his competitive ability. The climax is the loudest and highest-pitched phase of the call, often accompanied behaviorally by a short burst of locomotion, after which callers either stop vocalizing or produce a few, much quieter let-down elements.[34] Acoustic irregularities, or "nonlinear phenomena" (NLP), are more likely to emerge in these calls when vocal effort increases, such as occurs during the climax phase. NLP could function to enhance the auditory impact of calls, or they could simply reflect a caller's inability to maintain vocal stability. To examine these alternative hypotheses, climax calls were compared with screams, similarly loud and high-pitched vocalizations. Climax calls exhibited significantly fewer NLP than screams, and the harmonic components of climax calls that exhibited NLP had higher fundamental frequencies than climax calls without NLP. These results were interpreted to reflect effort by callers to produce stable, harmonically structured calls at the limit of their physiological ability, the outcome of this effort then potentially reflecting their physical vigor.[35] A subsequent analysis showed that males with higher mean annual and monthly testosterone levels began their climaxes at higher pitches. This finding supports the idea that the acoustic features of pant hoots might signal competitive ability.[36]

Group Membership

Pant hoots are audible for a kilometer or more and are commonly exchanged in intercommunity interactions.[37] Several studies have reported subtle population-level acoustic differences between pant hoots from different communities. These differences primarily concern the rates and pitch at which elements of the four call phases are produced, and they have been likened to differences in pronunciation rather than the phrase-level variation characteristic of songbird "dialects."[38] One study purported to show that the pant hoots of Kanyawara chimpanzees differed from those of Gombe and Mahale chimpanzees in frequently lacking a build-up phase, in a fashion analogous to dialectal variation in some bird species. However, a subsequent analysis comparing Ngogo and Mahale pant hoots failed to replicate the finding (recall that the Ngogo and Kanyawara communities are both located in Kibale National Park), and the researchers attributed the prior result to sampling error.[39] Whether slight geographic variation in pant hoots between communities is a consequence of habitat differences, body size differences, or vocal learning is unclear (discussed later).

Several studies have indicated that chimpanzees can distinguish the pant hoots of strangers and neighbors from those of their own community members, which would be advantageous in view of the potential for lethal interactions between communities (Chapter 8). In a playback study at Kanyawara, calls recorded from Mahale males were broadcast to parties of variable age/sex composition. Only parties with three or more males counter-called, exhibited signs of arousal, and approached the playback source (in one exception, a party with two males counter-called together with a third male a short distance away). By contrast, in two cases in which a pant hoot from a community male was broadcast, listeners exhibited no signs of fear, counter-called, and approached the source despite having fewer than three males.[40] In a playback study at Taï, where the communities tested only had two or three adult males each, listeners likewise tended to respond with screams to neighbor and stranger pant hoots but with pant hoots to calls recorded from members of their own community.[41] Consistent with these results, a follow-up study at Kanyawara showed that males pant hoot at lower rates when they are foraging at the periphery of their range, suggesting that their calls have the potential to evoke hostile reactions from neighbors.[42]

Multiple Messages

It is likely that receivers can glean multiple messages from some signals. For example, a pant hoot might encode individual identity as well as physical condition, or a bark might combine individual identity with the motivation to engage in aggression. Likewise, individual chimpanzees regularly produce more than one call type in succession or in combination during social interactions, reflecting multiple or changing motivational states. A subordinate individual, for example, may approach a higher-ranking one while alternating between screaming and pant grunting, indicating both fear and submission. Consistent with the dynamic nature of chimpanzee social interactions resulting from their fission–fusion social organization, some signals often occur in combination as individuals maintain auditory contact with group members out of view or adjust to renewed proximity after periods of separation.[43]

Information Content of Vocalizations 2:
"Functional Reference"?

Calls that are uniquely associated with a single environmental stimulus, such as food or a predator, could theoretically be used by listeners to gain information about phenomena external to the sender. The observation that receivers respond as if a vocalization refers to a feature of the environment can be taken as preliminary evidence that the call provides such information. Playback experiments may then generate further evidence that receivers are reacting to the call alone and not to additional contextual information. Such data, in turn, might suggest that the call refers to an external object in a fashion similar to the way a human word can. The resemblance, however, is only superficial.[44] In recognition of the fact that animal calls are fundamentally different than words, the concept of "functional reference" was developed to describe the production and consequences of some relatively context-specific animal vocalizations.[45] Straightforward operational criteria have been defined to identify functionally referential calls, although considerable debate persists regarding the theoretical soundness of the concept and its empirical justification.[46]

Chimpanzees frequently utter sequences of soft and loud, variably pitched grunt vocalizations upon arriving at preferred food sources and subsequently eating. Bouts of these intermingled "food grunts" and "food aaas" (hereafter called "food calls," collectively) can last for minutes, with brief pauses to listen to distant calls or simply to eat. The calls are short in duration, about 0.1 second, and bouts can include between 100 and 200 of them, though sequences are often shorter.[47] Because of the close association between feeding and calling, early observers suggested that the vocalizations might include a message about the discovery or presence of a favored food.[48] Anecdotal reports that listeners changed their travel direction and proceeded toward these food calls, and subsequently fed after arriving at food sources, provided additional support for this idea.[49] A later study conducted in captivity found that a juvenile male approached an artificial feeding station after recordings of food calls produced by his group mates were broadcast from the vicinity of the feeding location. The researchers in this study therefore concluded that these vocalizations should be considered "functionally referential."[50]

Despite the plausibility of the inference, there is little quantitative evidence that food calls encode a message about the location of food sources, the presence of individuals that are feeding, or the type of food being eaten. First, there have been no experimental playback studies in the wild examining the production or response specificity of these calls generally (endnote 46). Nor have the responses of listeners under natural conditions been systematically documented. A study at Taï reported that callers feeding in one species of fruiting tree were joined by more individuals when the first call of a bout was lower-pitched, but the behavior of the joiners prior to their arrival at the food tree was not monitored. It was therefore impossible to determine the effect of the calls alone on listeners.[51]

Second, there is also no compelling evidence that different food call variants function as labels for different food types, in a fashion analogous to the acoustically distinct alarm calls produced by some monkeys to different classes of predators.[52] A study at Budongo failed to demonstrate that chimpanzees produced different food grunt types when feeding on the fruits of different tree species.[53] In the captive study mentioned earlier, calls given to a preferred food type were higher-pitched and less noisy than those given to a less preferred food type. However, it was impossible to rule out the possibility that social factors were responsible for these acoustic differences. For example, because the sample of callers was small, individual, age, and sex differences in call structure were not controlled for. Moreover, increasing pitch in animal signals is associated with heightened arousal. Since the proposed food call types were acoustically very similar, differing primarily in frequency, they could simply have reflected differences in caller excitement.[54]

Third, and most importantly, efforts to discover whether chimpanzee food calls are functionally referential have relied on the analysis of only the very first call, or the first few calls, of calling bouts. Although methodologically convenient, this approach ignores the dynamic nature of food calling behavior. Over the course of extended bouts, the acoustic features of calls can vary constantly and dramatically. The subjective impression is that the pitch, loudness, and tonality of calls rise as vocalizers get more excited – for example, upon initiating feeding, when hearing other individuals calling and approaching, or when chorusing.[55] An anecdotal report from Gombe that points to the close connection between arousal and food calling is telling in this regard. In an apparent attempt to suppress food grunting and avoid attracting nearby high-ranking males after receiving bananas, a low-ranking male "made no loud sounds, but the calls could be heard deep in his throat, almost causing him to gag."[56] Studies examining how acoustic variation within bout sequences relates to contextual changes during feeding sessions are needed to explore alternative hypotheses for the functional significance of this acoustic signal (discussed later).

Vocal Behavior and Social Bonds

The selective production of some chimpanzee vocalizations appears to be indicative of close social relationships and in some cases may actively contribute to social bonding. For example, in a study at Budongo, males were more likely to give food calls if a preferred grooming partner was in the feeding party. In a follow-up study, recorded pant hoots were broadcast from a distance to males feeding silently in order to simulate the arrival of another male. Feeding males were more likely to respond with food calls if the recorded pant hoot was that of a close social partner. And finally, at Kanyawara, food calls were more likely and feeding bouts were longer when preferred social partners were in the party. Thus, food calling appears to reflect elevated arousal in the presence simultaneously of high-quality food and preferred group mates, and could theoretically promote group cohesion and bonding by signaling the likelihood of prolonged feeding.[57]

As already noted, pant hoots are produced by males in a variety of social contexts. In addition, two or more males often produce these calls simultaneously. Two studies at Mahale have demonstrated that when two males pant hoot together, their calls are more similar to each other acoustically than when they call separately or with other males. Subsequent studies at Kanyawara and Budongo indicated that the durations of pant hoot phases were longer when males were chorusing, potentially making it easier for callers to join in with one another. This suggests that some males actively modify their calls to facilitate chorusing and to match the calls of their partners. Since the chorusing males in these studies also associated and groomed together frequently, calling together appeared to reflect social bonding. Whether acoustic convergence reinforces existing social bonds, in a manner analogous to that suggested for other species, is unknown.[58]

AUDITORY COMMUNICATION 2: NONVOCAL SIGNALS

Buttress Drumming

Chimpanzees at all sites generate deeply resonant sounds that are audible for a kilometer or more by striking the buttresses or trunks of trees with their feet and, less often, hands. This "buttress drumming" typically takes the form of brief sequences of hits, on average three to six beats delivered in 0.5 to 1.5 seconds, although occasionally bouts are as long as four or five seconds.[59] Buttress drumming often occurs in conjunction with bursts of high-speed locomotion, either integrated into ongoing charging displays or simply as the culmination of a fast approach to a tree. Adult males produce the great majority of drumming bouts, drum most often during travel, and frequently do so together with pant hooting. For this reason, researchers have assumed that buttress drumming is a long-distant contact signal that functions to coordinate travel between separated chimpanzees and perhaps facilitate reunions.[60]

If chimpanzees had individually distinctive drumming styles, drumming bouts could be used by distant listeners to adjust their ranging behavior according to their motivation to associate with particular members of their community. Although it is unknown whether listeners in fact approach or avoid specific individuals in response to hearing them drum, a few studies have explored the possibility that males have drumming signatures. These efforts have focused on the number and timing of drum beats within bouts, and in particular on the common but variable inclusion of closely spaced pairs of beats separated by slightly longer pauses (referred to as the "double beat" pattern). Researchers at Taï and Budongo reported suggestive evidence of distinctive drumming signatures, while a study at Kanyawara failed to find such individual differences (endnote 59). The Kanyawara study also reported preliminary evidence for community-level differences in the integration of pant hoots and drumming, but this analysis was based on a very small sample of bouts from Taï. The possibility of intercommunity variation in drumming awaits further study.

It is unclear how preferences for particular drumming trees might contribute to systematic differences between the drumming displays of different individuals. Audiovisual analysis of a large number of drumming bouts from Gombe demonstrated that drummers generally approached and engaged trees to drum at a gallop, initiated bouts with gallop-gait limb sequences, and drummed primarily with their feet. Bouts were prolonged by holding onto the crest of a buttress with the hands while continuing to strike below with the feet as if galloping, or by leaping to additional buttresses to continue in like fashion (Plate 11). The double beat pattern observed in buttress drumming is a direct consequence of this locomotor footfall sequence and appears to reflect effort to deliver the most force to non-resonant substrates.[61] The inevitable consequence of this mode of production is that the size, shape, orientation, and spacing of buttresses will affect a drummer's ability to continue striking with hind couplets and will therefore influence the overall temporal pattern of beat sequences. It will be necessary to examine tree choices as well as angle of approach to assess the relative contributions of substrate geometry and individual motor skill to the timing of drumming bouts.

Other Nonvocal Auditory Signals

Two common nonvocal auditory signals produced in the context of social grooming have been briefly studied. The first, called "lip-smacking" and produced by the groomer, is a sound generated by opening and closing the mouth rapidly and clapping the lips together. Researchers at Budongo found that grooming bouts tended to be longer and were more likely to be reciprocated if lip-smacking occurred in the first ten seconds of a bout. Based on these results, they speculated that lip-smacking functioned to coordinate episodes of social grooming.[62] The second is a form of self-scratching that has been dubbed "directed scratch." It is distinguished from the self-scratching that occurs during periods of social excitement and is generally interpreted to reflect stress or frustration. By contrast, a brief study at Ngogo reported that self-scratching during grooming bouts was frequently followed by the groomer shifting his attention to the spot being scratched. Based on this observation, researchers speculated that scratching during grooming was an intentional signal that functioned to indicate which area of the body an individual wanted groomed.[63]

AUDITORY COMMUNICATION 3: POTENTIAL PRECURSORS OF LANGUAGE – VOCAL LEARNING, CONTROL, AND INTENTIONALITY

Tracing the evolution of language is challenging because the ancestral species whose members utilized premodern forms of speech have long been extinct, and fossil remains offer few clues about linguistic behavior. Nevertheless, humans are primates, and it seems reasonable to search for evolutionary precursors of language in the behavior of other primate species. One potentially informative avenue of research involves exploring the cognitive capacities that underpin complex social dynamics, capacities that may have been recruited by an evolving language capacity. A number of field studies on several Old World monkey species have addressed the question of language precursors from this perspective.[64] Additionally, language behavior involves more basic vocal capacities, and the possible antecedents of these general abilities have been explored through analyses of the development and use of vocal signals in a wide range of primate species. Research on chimpanzees in this area has included investigating the degree to which call structure is learned, whether individuals have voluntary control over vocal production, and whether individuals use calls intentionally to alter the behavior of their listeners.

Vocal Learning

Learning appears to have little influence on the acoustic structure of chimpanzee vocalizations, as is the case for vocalizations in other primates. As previously mentioned, several studies have revealed slight interpopulation variation in the pitch and rate of delivery of pant hoot elements. Such differences could reflect the influence of learning, but the effects of habitat acoustics and genetically based anatomical differences between individuals in different populations have been implicated as well.[65] Additional potential evidence of vocal learning comes from reports of captive chimpanzees that have adopted the call variants of newly introduced individuals or deployed unusual attention-getting sounds. However, the sounds involved in these captive studies have also been observed in wild chimpanzees.[66] In the absence of significant vocal learning, chimpanzees therefore appear restricted to a relatively limited, species-specific call repertoire.

Flexibility of Vocal Production

There is a close tie between the emotional or motivational state of callers and the call type they produce, an inference based on the fact that specific calls are regularly associated with specific types of behavioral contexts. In spite of this tight correspondence, however, the production of many calls is by no means reflexive or involuntary. This is most evident when individuals suppress call production. In a study at Kanyawara, females and low-ranking males refrained from pant hooting when foraging alone or in parties lacking high-ranking adult males, including when arriving at rich food sources, presumably to avoid inviting feeding competition.[67] A subsequent study at Kanyawara, as mentioned previously, found that when they were in parties containing fewer than three males, adult males refrained from pant hooting when hearing playbacks of stranger pant hoots. Adult males at Kanyawara also pant hooted less when ranging near their neighbors or when crop-raiding.[68] Finally, patrolling male chimpanzees remain unusually quiet (Chapter 8, endnote 11).

While call suppression is an obvious example of vocal flexibility, a series of studies at Budongo has also indicated that the production of several call types is affected by the composition of the caller's audience. In two studies of screaming, victims produced longer and noisier calls when high-ranking individuals were present, presumably in an effort to recruit their aid.[69] In studies of female copulation behavior, swollen females were more likely to produce copulation screams when mating with high-ranking males but suppressed calling when high-ranking females were in the vicinity.[70] In a study of submissive greetings, females were more likely to pant grunt to the alpha than to other males, and they were less likely to produce the call to other males when the alpha male was present compared with when he was absent.[71] And as mentioned previously, individuals were more likely to produce food grunts when preferred social partners were in the vicinity (endnote 51). Although the functional significance of specific audience effects remains speculative in all of this work, the analyses nonetheless point to a degree flexibility in chimpanzee vocal production and are consistent with evidence that chimpanzees have a sophisticated awareness of the social relationships of other group members (Chapter 1).

Intentionality

Finally, chimpanzees appear highly adept at reading the cues and signals of their group mates, and they give the impression of having some understanding of the mental states of other individuals. Nevertheless, there is as yet no evidence that wild chimpanzees communicate with the intent to provide ignorant receivers with information (i.e., second-order intentionality; see endnote 5).[72] Even in captivity, where numerous studies have been conducted under controlled conditions, evidence of mental state attribution remains inconclusive.[73] By contrast, there is some evidence from wild studies of first-order intentionality in the use of vocal signals, analogous to that found for some gestures (see "Intentional Gestures" earlier in this chapter). For example, in a study of soft "hoo" vocalizations given in the context of initiating travel, callers vocalized preferentially when allies were present, suggesting that the calls were directed at specific receivers.[74] In a second, experimental study at Budongo, individuals exposed to an artificial predator produced some alarm calls more often when "friends" were nearby and were more likely to stop calling when these individuals had moved away from the "threat." These results were interpreted to suggest that the calls were aimed at specific individuals and that callers monitored the receivers and persisted in calling until the goal of warning them was achieved.[75]

While there are plausible reasons to imagine that the gestural and vocal behavior of wild chimpanzees might be more complex than that of other nonhuman primates and consequently offer special insights into the evolution of language, there is little quantitative evidence to support this idea. Grounds for expecting that chimpanzees would exhibit superior abilities include the fact that they have comparatively large brains, are our nearest primate relatives, and live in a social context that potentially places special demands on individuals who must navigate relationships among intermittently seen community members. Moreover, a number of captive individuals have learned to associate hand gestures and artificial symbols with objects and to use them to communicate with humans. Study in the wild, however, has thus far failed to demonstrate that chimpanzee communicative behavior is especially sophisticated: The structures of calls and gestures are under strong genetic control, and signals appear primarily to contain information about senders. Still, forest conditions and fission–fusion social organization make it particularly difficult to conduct systematic research on wild chimpanzee communication. New observational and experimental techniques may yet demonstrate that the signaling behavior of these apes is more complex than current evidence indicates.

11

Community Differences in Grooming Postures and Tool Use: Innovation, Social Learning, and the Question of "Culture"

OVERVIEW

The social systems of wild chimpanzee communities appear fundamentally similar across the species range, with the exception of the very small Bossou group, which is hemmed in by human settlement and isolated from other chimpanzee populations. Nevertheless, there is a remarkable degree of variation between communities in the expression of numerous social and nonsocial behaviors. In some cases, the same general behavior may be performed in different ways in different communities. For example, distinctive postures adopted during bouts of social grooming are common in some populations but not others. Alternatively, a specific behavior, such as using stones to crack open nuts, may be exhibited in one population but not another, even though stones and nuts are available to both. Scrutiny of long-term records has revealed scores of behaviors that vary between field sites and many community-specific patterns that persist over time.

The existence of persistent, community-level differences in behavior raises a host of questions about chimpanzee cognition. Answers to these questions will greatly improve our understanding of chimpanzee behavioral flexibility and will contribute to hypotheses about the evolution of cultural behavior in human ancestors. For example, to what extent are behavioral variants socially learned? If social learning is implicated, what types of learning are involved? Do chimpanzees teach each other skills? What conditions favor innovation? Are some individuals more likely to be innovators than others? Do behavioral innovations build on each other? How important is inter-individual tolerance for the transmission of behavioral variants? Do behavioral patterns diffuse across populations? Are chimpanzees inclined to conform to group patterns? And finally, do chimpanzees experience a sense of group identity as a consequence of shared patterns of behavior?

Such questions about chimpanzee cognition, however, are difficult to investigate under field conditions. There are several reasons for this. First, it is difficult to disentangle the various influences between infancy to adulthood that may contribute to the development of a specific behavior in an individual. Second, the spread of a novel behavior, introduced through either spontaneous innovation or immigration, is a comparatively rare occurrence. Finally, if a researcher is lucky enough to observe a newly introduced behavior, analyzing its potential diffusion then requires comprehensive and long-term study that is not easily accomplished under field conditions. As a consequence, research exploring precisely how chimpanzees acquire new behaviors and how new behaviors may diffuse across groups is largely conducted experimentally on captive animals.

It is beyond the scope of this chapter to review the extensive and rapidly growing body of research on chimpanzee cognition and learning carried out in captivity. Instead, a few key results are noted along with a more general review of intergroup behavioral variation in the wild, which has been documented primarily in the context of tool use. Debate over the potential relevance of intergroup variation among chimpanzees for understanding the evolution of human cultural behavior is also briefly considered.

IDENTIFYING COMMUNITY-SPECIFIC BEHAVIORS INFLUENCED BY SOCIAL LEARNING

Methodological Considerations

In a pioneering collaborative effort, researchers at nine long-term study sites identified thirty-nine "[behavior] patterns customary or habitual at some sites yet absent at others," the majority of which involving tool use.[1] The authors considered it likely for two reasons that such community-specific behaviors were established and perpetuated through some form of social learning: The differences could not be explained by variation in ecological conditions, and they were present between communities of the same subspecies and therefore unlikely to be a consequence of genetic differences.[2] The geographical distributions of distinctive behavior patterns were also mapped in order to investigate three possible mechanisms of diffusion from communities of origin: (1) from a unitary source to successive neighboring communities, (2) from multiple sources (i.e., independent innovations) to neighboring communities, and (3) from one or more sources to neighboring communities while being modified slightly in the process ("diffusion with differentiation"). The distributions of several community-specific behaviors were consistent with each of these three diffusion models.

The inference that social learning is involved in the establishment and perpetuation of behavior patterns within communities is supported by detailed observations of the development of tool-use behaviors by young chimpanzees. For example, beginning as early as six months of age, infants watch intently, often at very close range, as their mothers or other adults crack open nuts using "hammers" and "anvils" or fish for termites with plant stems (Plate 12). The percentage of time during tool-use sessions that infants attend to these behaviors increases as they get older. In the case of termite fishing, infant females spend more time than infant males watching their mothers and later are more proficient in the behavior.[3] By roughly one and a half years of age, youngsters begin to examine and manipulate discarded tools without trying to deploy them. Mothers are extremely tolerant of their offspring during these foraging episodes and may allow them to take and use their tools. By roughly two and three years of age, respectively, infants attempt to use them to harvest termites and crack nuts. Their early efforts are conspicuously clumsy and inefficient, but with practice their proficiency improves, as measured by the rate at which they extract prey items.

Research in captivity also supports the conclusion that social learning is involved in the establishment of enduring, community-level behavioral variation. Innovative experiments have demonstrated that tool techniques introduced to isolated chimpanzees will spread faithfully to their group mates after the knowledgeable subject is reunited with its fellows.[4] Moreover, once seeded and transmitted within one group, the behaviors will subsequently diffuse to new groups that have been permitted to observe the initial subjects.[5] In the non-tool-use domain, researchers have investigated whether a self-medicative behavior common among wild chimpanzees will spread once introduced into a captive group. Wild chimpanzees swallow the leaves of several plant species whole and subsequently defecate them undigested several hours later, a behavior associated with the expulsion of gut parasites. After a few individuals first exhibited the behavior in two captive groups, group-level differences in the method of ingestion emerged over time, with individuals biased toward the method of those who first exhibited the practice in their group (see also later).[6]

Nevertheless, despite the likelihood that intercommunity behavioral variation is established and maintained through social learning, it has been difficult to demonstrate this unequivocally in the wild. First, although rapid, short-term diffusion of novel behaviors has been reported, cases of an observed innovation spreading and then becoming widespread and enduring in a population are rare.[7] Second, it is difficult to evaluate possible microecological influences on local behavior patterns. For example, differences between communities in tool choice and method of harvesting army ants may partly result from differences in prey characteristics between field sites.[8] Third, some variation may result from individual predispositions for independently discovering a behavior. In the whole-leaf swallowing experiments mentioned previously, groups of captive chimpanzees who had never seen the behavior were initially provided with leaves that resembled in texture those swallowed whole in the wild, and several individuals in each group spontaneously exhibited the behavior.

Variation in Tool Use between Communities

Tool use in animals can be defined as "the use of an external object as a functional extension of mouth or beak, hand or claw, in the attainment of an immediate goal."[9] Although tool manufacture and use are widespread in the animal kingdom, the breadth of activities for which primates exhibit them is unparalleled.[10] And among the tool-using primates, chimpanzees are exceptional. Individuals in some communities use stones to break open nuts, and chimpanzees everywhere employ all manner of plant parts (leaves, stems, twigs, small and large sticks, branches, saplings, etc.) for various probing and investigative activities, harvesting insects and honey, soaking up water to drink, self-cleaning, shooing flies, threatening and attacking other individuals, generating acoustic and visual signals, and protecting body parts.[11] New tools and behaviors continue to be documented, both at long-term study sites and at newer ones.

Many of the tool-use behaviors displayed by chimpanzees are relatively simple and widespread among primates, such as shaking vegetation to threaten group members. Others, however, are considerably more complex, revealing an understanding of a broad range of materials in the environment and requiring foresight and planning.

For example, peeling leaves off of a stem to fashion a probe with which to collect ants from a nest displays an appreciation of the functional potential of the plant part, but one that can only be actualized after modification of the original material. Still greater complexity is exhibited in the use of multiple tools to perform a task – for example, deploying a "hammer" and "anvil" to crack a nut while bracing the anvil with one or two additional stone wedges. And complexity of yet a different sort is demonstrated in the sequential use of different tools to achieve a goal. For example, up to five objects may be used to harvest honey, with distinct implements to whisk away bees, create and then enlarge an opening in the hive, and finally extract the honey.[12]

The creativity with which chimpanzees fashion and use tools inevitably results in individual differences in behavior, some of which rise to the level of community-wide patterns. Of the thirty-nine behavior patterns identified in the pioneering study described earlier, thirty-four can be obviously described as involving tool use. Three more could be considered tool use in the broadest sense defined previously, that is, involving the use of an external object (two versions of pounding food on something hard, and branch shaking or buttress beating in charging displays performed during heavy rain). Roughly half of the behaviors exhibiting community-level variation occurred in the context of investigating or acquiring food. Only differences in grooming postures, along with a method of killing insects detected while grooming, were clearly non-tool-use activities, although several other non-tool-use behaviors have since been tentatively identified as occurring in some populations but not others.[13]

Table 11.1 provides a tally of the number of behaviors that were found in the initial collaborative study to be customary or habitual in some populations but absent in others, broken down by the type of activity involved.[14] Note that twenty-six mostly tool-use behaviors were either shared by all communities, not habitual anywhere, or their absence was explained by some environmental factor.

Variation in Grooming Behaviors between and within Communities

Chimpanzees can assume any number of postures while grooming and being groomed, depending on the part of the body being groomed and on whether the groomers are on the ground or in the canopy. Typically,

Table 11.1 *Number of behavior patterns customary or habitual at some sites but absent at others, grouped in terms of activity type involved[1]*

Behavior type	Total subsistence-related	Total other types	Numerical entries listed in Whiten et al. (1999, 2001)[2]
Pounding actions	9		27–35
"Fishing" for insects	5		36–40
Probing	3		41–43
Forcing	2		44–45
Leaf seat, fly whisk		2	46–47
Self-tickle with object, throwing		2	48–49
Clean body, dab wound with leaves		2	50–51
Strike or manipulate vegetation to generate acoustic or visual signal[3]		9	52–55, 60–64
Use leaf to squash ectoparasite		1	56
Use leaf to inspect ectoparasite		1	57
Kill ectoparasite with forefinger hit		1	58
Hand-clasp grooming posture		1	59
"Rain dance"[4]		1	65
Total	19	20	

[1] Derived from Whiten et al. (1999, 2001), who compared long-term records from nine study sites. They used the following definitions (2001: 1488): Customary, "pattern occurs in all or most able-bodied members of at least one age-sex class (*e.g.* adult males)"; habitual: "pattern is not customary but has been seen repeatedly in several individuals, consistent with some degree of social transmission." All of the behaviors involved tool use, with the exception of hand-clasp groom, kill ectoparasite with forefinger, and rain dance. Refer to Whiten et al. (2001) for definitions of specific behavior patterns.

[2] Whiten et al. (1999, 2001) numbered the sixty-five behaviors they examined. Of entries 1–26, 1–8 were "patterns absent at no site"; 8–23 were "patterns not achieving habitual frequencies at any site"; and 24–26 were "patterns for which any absence can be explained by ecological factors" (1999: Table 1).

[3] Note that although as originally described, "leaf groom" (#52) appeared primarily communicative in nature, observations from Mahale suggest that sometimes at least it is performed to squash ectoparasites, as in #56 (Zamma, 2002).

[4] Absent only at Bossou.

both groomer and groomee sit, although an especially relaxed individual may recline partially prone, resting on a hand or elbow, or lie prostrate while being groomed. Most of the time only one individual grooms the other (unidirectional) or the two take turns (reciprocal or bidirectional). Less frequently, although commonly between at least some pairs, the two individuals groom each other simultaneously, a behavior referred to as mutual grooming.[15] As in other forms of grooming, participants position and reposition themselves during mutual grooming to expose their own body parts or to access different body parts of their partners. In one distinctive pattern, customary or habitual in all communities, the two individuals each have an opposite arm held overhead grasping a branch or vine (i.e., both right arms are up or both left arms are up) as they simultaneously groom under each other's raised arm.[16]

A particularly striking configuration of mutual grooming with raised arms, originally labeled the "grooming-hand-clasp," occurs when "each of the participants simultaneously extends an arm overhead and then either one clasps the other's wrist or hand, or both clasp each other's hand."[17] The behavior is comparatively brief and occurs at the beginning of or intermittently during grooming bouts.[18] Since its initial description, this grooming style has been reported to be customary or habitual in nine communities and absent in three.[19] In addition, since it can take a number of different forms that do not always include holding hands, the more general term "high-arm grooming" has been proposed to replace "grooming-hand-clasp" or "'hand-clasp grooming,'" with subtypes then defined by the nature of the bodily contact: palm-to-palm clasp, wrist-to-wrist, palm-to-wrist, arm-to-arm, and so on (Plates 13 and 14).[20] In addition to intercommunity variation in the occurrence of high-arm grooming generally, variation also exists within and between communities in the proportion of different subtypes exhibited by individuals.[21]

Several other behaviors associated with grooming also vary between field sites. Individuals in some communities occasionally grab leaves, peer at them closely, and then begin to manipulate them, a behavior referred to as "leaf grooming." This may function as a communicative signal, either encouraging a current partner to continue or attracting others to groom.[22] Alternatively, leaf grooming may involve active searching for insects.[23] Leaves may also be used to inspect or smash ectoparasites, and characteristic hitting motions are sometimes deployed to kill insects placed on the palm or forearm. Finally, "social scratch" is a behavior that also occurs during social grooming, in which an individual "rakes the hand back and forth across the body of another" in a fashion similar to when scratching themselves.[24] It was first described at Mahale and has since been observed at Gombe and Ngogo, with some variation: At Ngogo and sometimes at Mahale, it is executed with fingers held straighter and the motion more like poking.[25]

THE CULTURE DEBATE: INNOVATION, SOCIAL LEARNING, AND THE QUESTION OF CONFORMITY

Community-specific behavioral variation in chimpanzees is reminiscent of observable differences in behavior between modern human groups, differences that we commonly label as "cultural." Since the 1970s, chimpanzee researchers have likewise labeled persistent, community-level differences in chimpanzee behavior as "cultural" and referred to distinct constellations of local behavior patterns as chimpanzee "cultures."[26] A widely cited rationale for using this terminology is that locale-specific behavior patterns in chimpanzees satisfy a set of criteria distilled from an early twentieth-century effort to describe human culture as "a relatively closed system of phenomena": namely, that behavior patterns reflect innovation (something new is invented), dissemination (gets passed on), standardization (is consistent across individuals), durability (is performed without being in the presence of the demonstrator), diffusion (spreads between groups), and tradition (is transmitted to the next generation).[27] Fundamental to this concept of culture is that some mechanism of social learning is implicated in the acquisition of distinctive behavior patterns by individuals.[28]

Yet if the designation "culture" is merited whenever animals exhibit persistent group-level behavioral variation mediated by some form of social learning, then a wide range of species qualify as "cultural" animals, from birds to rodents to marine mammals.[29] Since the concept of culture was developed to describe apparently unique aspects of human group behavior, applying it to other animals minimizes the conspicuous gulf that exists between human and nonhuman sociality. A compelling argument can be made, therefore, to restrict the concept of culture to humans, since it encompasses much more than the transmission of behavioral techniques. For example, whereas animals may pass along a few, relatively simple behavior patterns, human cultural practices are modified over time, with modifications accumulating across generations (the so-called "ratchet effect").[30] Humans also learn and perpetuate moral rules, rituals, and social institutions that regulate social life and contribute to a conscious sense of group identity.[31] Lacking such features, chimpanzee intergroup behavioral variation, from this perspective, would merely reflect an exceptional capacity for innovation and social learning.[32]

Nevertheless, even in the absence of a cumulative ratchet effect or the symbolic regulation of group behavior, it is conceivable that to

some degree individual chimpanzees recognize and attribute significance to community-level behavioral variation. Faced with the hostility
of intergroup interactions (Chapter 8), chimpanzees clearly maintain
an acute awareness of group membership. Is it then possible that
individuals experience a sense of group membership based not just
on firsthand knowledge of who their group mates are but also on
distinctive behavior patterns exhibited by a majority of their community members? If so, this would suggest that they exhibit the glimmer-
ings of an important feature of human cultural life. To explore this
possibility, researchers have therefore examined whether chimpanzees conform to group norms, in the simple sense of matching their
behavior to that exhibited by the majority of individuals in their group.

As described previously, tool-use techniques introduced to captive individuals will spread to other group members, and different
techniques can come to predominate in different groups. These experiments have been interpreted to provide evidence that the chimpanzees
conformed to the patterns exhibited by the majority of their compa-
nions, since individuals that discovered alternative techniques nonetheless continued to deploy those of their group mates.[33] Similarly,
researchers at Taï documented slight but long-term differences in nut-
cracking hammer selection (whether stone or wood and what size)
between Taï communities despite the migration of females between
groups. A case study of one immigrant female indicated that she gradually abandoned the hammer material selected by most individuals in
her natal community and increasingly selected the hammer type most
commonly used in her new community. This shift was likewise proposed to reflect conformity to a group norm.[34]

Despite such speculative interpretations, however, there is reason to suspect that rather than adjusting their behavior to group
norms, chimpanzees simply adopt behaviors similar to those of the
individual demonstrators they observe. In support of this idea, behavioral convergence within captive groups can occur prior to the existence of a majority pattern. Various studies also indicate that
chimpanzees tend to be highly resistant to changing their behavioral
strategies and frequently persist in using their first-learned technique
for a given task unless a more profitable alternative is discovered.[35]
Moreover, studies in the wild and in captivity consistently indicate that
behavioral variants spread preferentially through matrilines, which
emphasizes the importance of the mother's influence in ontogeny,
and between the most affiliative unrelated individuals.[36] These observations suggest that chimpanzees have a strong tendency to be

influenced by socially acquired information but not necessarily by majority practices.[37]

In conclusion, chimpanzees combine a long developmental period with a considerable capacity for social learning and a propensity to try new behaviors. These characteristics contribute to an unusual degree of behavioral variation between communities in tool-use and social behavior. To date, however, convincing evidence is lacking for a ratchet effect in group behavior patterns, symbolic mediation of group behavior, or conformity to group norms, all of which are routine features of human cultural life. This suggests that referring to persistent community-level differences in chimpanzee behavior as "cultural" fosters a misleading perception of their significance, insofar as "culture" is associated with unique features of human group behavior. Nevertheless, analyses of chimpanzee intercommunity variation continue to provide insights into behavioral innovation and mechanisms of social learning. And just as ongoing analyses of signaling hold out the promise of discovering subtle complexities in chimpanzee communicative behavior, studies of intercommunity behavioral variation may well provide indications that individuals attribute greater significance to community-specific behavior patterns than has thus far been apparent, and contribute to ongoing efforts to understand the emergence of cultural behavior in human ancestors.

Epilogue

There is something disconcerting about watching chimpanzees in the wild. Whereas other African mammals simply seem like part of the natural landscape, chimpanzees can seem a little too big and resemble humans a little too much to be in the trees. You get the feeling while observing them that their primary food, ripe fruit, is inadequate to sustain them, perhaps because you cannot imagine eating so much fruit yourself, or perhaps because they seem too intelligent to sit around gorging on figs for hours on end. Or maybe it is their obvious and familiar social awareness and behavioral individuality that engender the impression that they are just slightly disconnected from the seamlessness of the environment around them. Or the fleeting but human-like social gestures that are invariably captivating – a light touch after a fight, juveniles chasing each other in play, a mother grooming her infant. Whatever the reasons, wild chimpanzees can seem to occupy an ambiguous place in the rain forest, necessarily part of their ecological communities yet somehow apart.

The traces of familiarity that ignite our visceral fascination with chimpanzees are also the foundation of our intense scientific interest in them. Chimpanzees and their sister species the bonobos are our closest genetic relatives, with whom we shared a common ancestor some 6 to 9 million years ago. Chimpanzees, in particular, exhibit what appear to be precursors of many of the behaviors long thought to make humans special among the primates: complex tool fabrication and use, reliance on hunting and meat eating, cooperative lethal aggression, language, and cultural behavior. Consequently, human evolutionary biologists draw on the behavior of chimpanzees to develop hypotheses about the evolution of these traits in our ancestors. Chimpanzees offer a jumping-off point for thinking about the kinds of changes that would have been necessary to propel our ancestors on the trajectory toward *Homo sapiens*. As Jane Goodall has remarked,

> ... it is only through a real understanding of the ways in which chimpanzees and men show *similarities* in behavior that we can reflect with meaning on the ways in which men and chimpanzees *differ*. And only then can we really begin to appreciate, in a biological and spiritual manner, the full extent of man's uniqueness.[1]

But in studying these apes in their natural habitats, we are necessarily confronted with the plights of their communities, and of the human communities that surround them. While this book focuses rather narrowly on chimpanzee behavior, fundamental to its purpose is the idea that long-term research projects have the potential to become critical loci of conservation efforts. It is impossible to study wild chimpanzees and not think about the threats to their survival and the importance of countering those threats. Deforestation, capture for sale as pets or zoo specimens, hunting for bushmeat, and death from human-borne diseases are the worst direct conservation menaces for chimpanzees at present. More broadly, however, chimpanzee conservation is a global problem, to the extent that local threats are linked to wider forces driving unsustainable resource extraction, economic inequality and political instability in host countries indelibly marked by European colonialism, and regional conflicts that force mass movements of human refugees. Thus, chimpanzee conservation involves not just enacting and enforcing local laws to protect forests and wildlife, but also addressing grave political and economic problems both near and far.

My intention has been to present a generally accessible overview of wild chimpanzee social behavior, and to provide a bibliographic resource useful to chimpanzee and other animal behavior researchers. But preserving viable chimpanzee populations in the wild depends only partly on the work of scientists. Ambitious educators, community development workers, legal experts, and policy makers, to name a few, also have key roles to play. My hope is that this book will also inspire engagement, whatever one's area of interest or expertise.

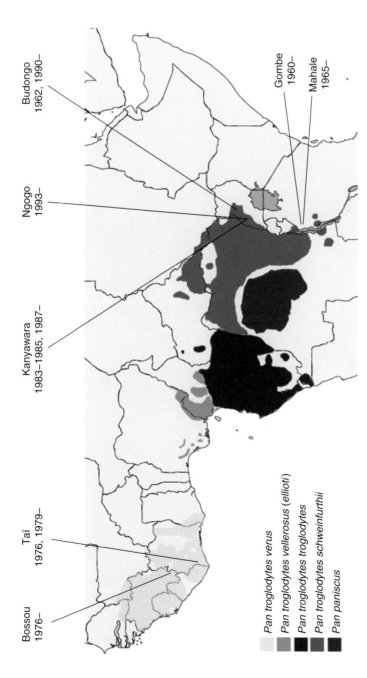

Fig. 2.1. Locations of the seven long-term field studies that have generated the most data on wild chimpanzee social behavior. See text for details

Bossou
1976–

Taï
1976, 1979–

Kanyawara
1983–1985, 1987–

Ngogo
1993–

Budongo
1962, 1990–

Gombe
1960–

Mahale
1965–

Pan troglodytes verus
Pan troglodytes vellerosus (ellioti)
Pan troglodytes troglodytes
Pan troglodytes schweinfurthii
Pan paniscus

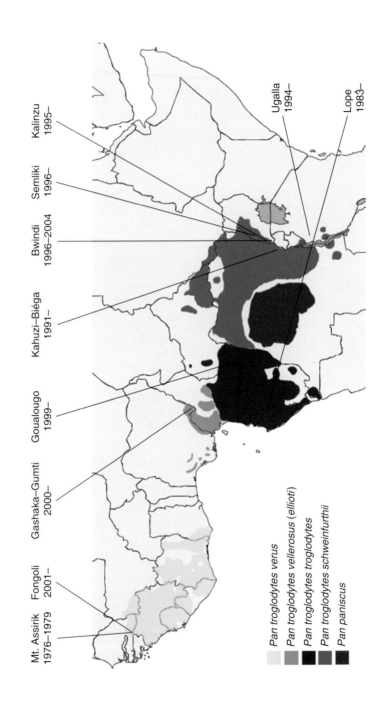

Fig. 2.2. Locations of some other important field studies

Mt. Assirik 1976–1979
Fongoli 2001–
Gashaka–Gumti 2000–
Goualougo 1999–
Kahuzi–Biéga 1991–
Bwindi 1996–2004
Semliki 1996–
Kalinzu 1995–
Ugalla 1994–
Lope 1983–

Pan troglodytes verus
Pan troglodytes vellerosus (ellioti)
Pan troglodytes troglodytes
Pan troglodytes schweinfurthii
Pan paniscus

Plate 1. Party of Kanyawara chimpanzees feeding on ripe figs (*Ficus capensis*). Ripe fruits are the mainstay of the chimpanzee diet (Chapter 1). (Courtesy of Ronan Donovan/Kibale Chimpanzee Project.)

Plate 2. Adult male Twig, who at age five caught his hand in a wire snare set by poachers to catch game animals. Chimpanzees are usually unable to remove snares, which cut off circulation to the appendage and cause necrosis. The appendage ultimately falls off and the wound may heal (Chapter 2). (Courtesy of Andrew Bernard/Kibale Chimpanzee Project.)

Plate 3. Family group resting and grooming. From right to left: 12-year-old son Tuber grooms his mother Tongo, who grooms her 7-year-old daughter Tsunami, with infant female Thatcher between them. Chimpanzees live in communities that exhibit 'fission-fusion' social organization, meaning that community members do not collect in a single, spatially cohesive group, but instead associate in small subgroups ('parties') separated throughout the community range (Chapter 3). (Courtesy of Andrew Bernard/Kibale Chimpanzee Project.)

Plate 4. A party traveling and foraging. Three adult males are followed by a family group: 8-year-old male Likizo, his mother Lia with 3-year-old daughter Azania riding on her back, and 13-year-old male Tuke pausing to eat. Ranging and association patterns are influenced by food availability and differ between males and females (Chapter 4). (Courtesy of Andrew Bernard/Kibale Chimpanzee Project.)

Plate 5. Tongo resting with her two daughters, infant Thatcher and 7-year-old Tsunami. Adult female life is focused around offspring care (Chapter 5). (Courtesy of Andrew Bernard/Kibale Chimpanzee Project)

Plate 6. Adult female Michelle, with her infant clinging ventrally, in an aggressive interaction with Tripoli, a young immigrant female (note the clinging infant's right foot grasping the mother's hip). Resident females are often aggressive toward new immigrants, who compete with them for food resources (Chapter 5). (Courtesy of Andrew Bernard/Kibale Chimpanzee Project.)

Plate 7. Three-year-old male Moon inspecting adult female Leona's estrous swelling. Ovulation occurs toward the end of maximal swelling, and consequently swelling is a cue to a female's approximate reproductive condition (Chapter 7). (Courtesy of Andrew Bernard/Kibale Chimpanzee Project.)

Plate 8. Adult males on patrol at Ngogo. Males regularly monitor community boundary areas, and will attack and sometimes kill neighbors that they encounter (Chapter 8). (Courtesy of John Mitani/Ngogo Chimpanzee Project.)

Plate 9. Begging cluster around a recently killed red colobus monkey. Chimpanzees are avid hunters, and share meat after captures (Chapter 9). (Courtesy of Ronan Donovan/Kibale Chimpanzee Project.)

Plate 10. Adult male Kakama exhibits a 'hoot face' while producing a long-distance 'pant hoot' vocalization. Chimpanzees use a wide range of facial, gestural, and vocal signals in social interactions (Chapter 10). (Courtesy of Ronan Donovan/Kibale Chimpanzee Project.)

Plate 11. Adult male Kakama 'buttress drumming,' which generates low frequency sounds audible for a kilometer or more. Note that the 'beats' are primarily produced by holding the crest of the buttress with the hands while slamming the feet into the wall below, typically beginning with footfall timing identical to that of galloping (Chapter 10). (Courtesy of Ronan Donovan/Kibale Chimpanzee Project.)

Plate 12. 3-year-old Azania watches her mother Lia dip for honey with a twig that she has peeled the leaves off of. Chimpanzees learn to fashion and use more tools for more activities than any other animal besides humans. Tool 'kits' also vary between communities (Chapter 11). (Courtesy of Andrew Bernard/Kibale Chimpanzee Project.)

Plate 13. Adult female Rosa, with her 2-year-old daughter Betty behind her, palm-to-palm 'high-arm grooming' with adult male Kakama. High-arm grooming occurs in some communities but not others, and different forms of high-arm grooming appear to be passed on from mothers to offspring (see Plate 14 and Chapter 11). (Courtesy of Andrew Bernard/Kibale Chimpanzee Project.)

Plate 14. Adult males Twig and Kakama engaged in wrist-to-wrist high-arm grooming (see Plate 13 and Chapter 11). (Courtesy of Andrew Bernard/Kibale Chimpanzee Project.)

Appendix: Field Methods for Studying Wild Chimpanzees

HABITUATING, FINDING, AND FOLLOWING CHIMPANZEES

Wild chimpanzees flee from people if they are unaccustomed to seeing them. If they are tracked persistently and unobtrusively, however, they will gradually become used to the presence of human observers. This process of "habituation," in which the animals increasingly ignore unobtrusive onlookers, progresses through several stages. In the early phases, individuals feeding in the safety of the forest canopy will tolerate being watched from below but will quickly run once they descend to the ground to travel. Some months later, they may tolerate being quietly followed as they travel on the ground to a new feeding area, provided the observers remain twenty or thirty meters away. Some individuals may even rest and socialize on the ground in view of onlookers, as long as the humans keep their distance. As habituation continues, some animals begin to come within five to ten meters, perhaps passing by a researcher to approach a grooming partner, but will still move away if the researcher tries to approach them to that distance. Finally, after years of being tracked, most chimpanzees appear to ignore observers completely as they pursue their daily activities.[1] At this stage, now fully habituated, they may literally brush past a researcher as they make their way along a path.

The most desirable data on wild chimpanzees are collected from fully habituated populations that are otherwise largely shielded from contact with humans. There are several reasons for this. First, researchers do not want their presence to affect the behavior of the animals they are watching. This appears to be achieved after complete habituation, when study animals show no visible fear of humans and engage in routine social and subsistence behaviors without looking at or otherwise directing behavior toward observers. Although chimpanzees everywhere have probably been influenced to some degree by human contact (people enter forests regularly for a variety of reasons – to collect firewood, edible plants, and medicinal plants; to hunt game, including chimpanzees at some locations; to extract lumber; to find refuge during warfare; and so on), habituated animals in protected areas nevertheless appear to offer an accurate picture of chimpanzee life unaltered by human influence.

Second, habituation ensures that researchers can more quickly and reliably identify their study animals. Individual chimpanzees are quite distinctive physically, and observers can learn to tell them apart within a few months based on physical characteristics such as size, hair and skin color, facial features, scars and deformations from injuries, and gait patterns. Ultimately, experienced observers can identify individuals instantly from a gestalt of features, as we do with humans. This ability is critical when collecting data during fast-paced social interactions that involve several group members, lots of movement, and often difficult observation conditions due to low light and obstructive vegetation. It is therefore essential to be able to identify each animal at a glance, and this is most easily achieved at close range with habituated subjects.

Third, complete habituation means that all members of a study population can be included in predetermined sampling protocols, producing a more accurate picture of group life. Invariably during the habituation process, some individuals prove to be less skittish than others and are habituated more quickly. Data collection during the early phases of a field study will consequently overrepresent these animals. In time, less habituated individuals gradually lose their fear, and observers can follow them more often. With complete habituation of the study group, researchers can build balanced data sets to test hypotheses about the functional significance of the behaviors they are observing.

Finally, once individual chimpanzees are fully habituated, researchers can more regularly observe them throughout the day, conducting so-called "all-day follows." This is because habituated chimpanzees can be tracked continuously until they build and settle down in their night nests (Chapter 1), and then found again at daybreak before they set out to forage. Prior to complete habituation, when observers can easily frighten the animals and must follow them from further away, it is common to lose track of them in dense habitat. If observers do not know where their targets nest for the night, their ability to find them the following day is severely hampered. In this event, they must then search randomly in the forest listening for loud calls or buttress drumming, visit trees with ripe fruits that the chimpanzees have recently been feeding in, or visit other trees known to be in fruit and that might attract their study animals. Individual chimpanzees may even be lost for days or weeks at a time if they forage and move about quietly.

HYPOTHESIS TESTING AND THE NEED FOR STANDARDIZED OBSERVATION PROTOCOLS

Scientific study of wild chimpanzee behavior depends on the systematic collection of empirical data. In general, researchers monitor the activities of known individuals for predetermined amounts of time, following standardized data collection protocols and using widely shared definitions of chimpanzee social behaviors. Data are typically recorded on check sheets, either by hand or on portable computers, or dictated and later transcribed. Sound and audiovisual data are also collected in some types of studies. Quantifiable data are then analyzed statistically to identify key variables within and between populations in order to develop and test hypotheses about the adaptive significance of social behaviors. Anecdotal information may supplement quantitative data, most often to suggest future lines of inquiry for which systematic information is currently lacking.

There are many practical challenges to collecting reliable and useful data under field conditions, making strict data collection procedures an essential feature of successful research. First, observation contexts are constantly changing in natural environments: the types, locations, and distribution of foods; the presence of predators and/or competitors; the variable presence of group mates; the sexual state of females; individual moods and histories of recent events such as fights or coalitions; the weather; and on and on. Thus, unlike chemists or physicists who can exert control over experimental conditions, field scientists typically rely on multivariate statistical analyses of long-term records to identify and rank causes of variation and, hopefully, illuminate the functional significance of behaviors. Carefully designed protocols become the foundation of successful efforts to collect data amenable to such analyses.

Second, systematic sampling methods are critical because there is usually too much behavior occurring at any given moment in a group of social animals for an observer to keep track of at once. Faced with information overload, field researchers can only record a fraction of what they see. This in turn creates the challenge of collecting data that will produce a faithful picture of overall behavior patterns. Standardized sampling procedures help observers restrict their attention to particular behaviors of interest and distribute data collection evenly across contexts and classes of individuals, increasing the likelihood that their observations will reflect reality accurately. For example, if the researcher wants to know whether males produce a particular vocalization more than females, she needs to collect comparable amounts of data over comparably long periods of time under a comparable range of conditions from all males and all females. Otherwise, she might inadvertently oversample an especially noisy individual or oversample a context more likely to elicit the call and then overestimate the frequency at which the call is uttered by males or by females. Researchers therefore rely on predetermined sampling rules that randomize which individuals are observed and standardize observation intervals to minimize bias in their data collection.

Finally, standardization of data collection protocols increases the opportunity for investigators at different study sites, and for different investigators working at the same study site at different times, to conduct collaborative and comparative research. Chimpanzees exhibit substantial behavioral complexity and flexibility, the full extent of which can only be documented if researchers study as many genetically diverse populations as possible, in the full spectrum of habitats. However, an accurate picture of behavioral variation across the species range requires agreement on definitions and sampling methods. For example, to understand what conditions prompt chimpanzees to hunt monkeys, researchers need to agree on an operational definition of hunting. Is a chimpanzee hunting when it simply looks up at a troop of red colobus monkeys in the canopy? Does it have to be climbing toward them? Running toward them? Is a chimpanzee hunting if he is following on the ground below while other chimpanzees are giving chase in the canopy? Clearly, the significance of reported differences in hunting behavior between different studies will depend on the criteria researchers use to identify and document the behavior.

BEHAVIORAL EVENTS VERSUS BEHAVIORAL STATES

Behaviors can be divided into two classes, corresponding to the opposite ends of a spectrum of durations. Brief behaviors lasting seconds or minutes are referred to as behavioral "events." Examples of events might include producing a vocalization, chasing an individual from a feeding location, or copulating. Longer-duration behaviors, lasting from minutes to hours, are called behavioral "states." Examples of states include feeding, walking, and grooming. In general, it makes the most sense to think about events in terms of the frequency at which they occur (e.g., events/hour) and states in terms of the proportion of waking time they represent.

The importance of the distinction between events and states becomes clear when one compares analyses of frequency and analyses of proportion of waking time for each class of behavior. For example, imagine a vocalization that lasts 0.4 seconds, and females produce an average of three of these vocalizations in a day, whereas males produce an average of twenty-one per day. Treated as an event, we might find that there is a large difference in calling frequency (calls/hour) between males and females. We might then discover that this difference is correlated with important differences in social behavior. By contrast, treated as a state, 1.2 total seconds of calling versus 8.4 total seconds of calling in twelve waking hours would seem rather unimportant and suggest (misleadingly) that sex differences in sociality do not influence communicative behavior.

Similarly, suppose we wanted to know whether male chimpanzees groomed with other males more than with females, and we observed males grooming with males and females about the same number of times per day. However, male–male grooming sessions lasted an average of ten minutes without interruption, whereas male–female grooming sessions lasted only thirty seconds. Treated as an event, we could say that males groomed with females as often as they groomed with other males, implying that female–male bonds were similar in strength to male–male bonds. But if grooming were treated as a state, we would have to conclude that males groomed with males much more than they groomed with females. From this information, we might then go on to infer that male–male bonds were much stronger than female–male bonds, laying the groundwork to develop hypotheses about the functional significance of bonding between different classes of individuals in the maintenance of group social organization.

SAMPLING METHODS

The choice of methods for any animal behavior study depends on the type of information needed and the feasibility of collecting it in an objective way. In some circumstances, several methods can be deployed simultaneously. For any protocol, the minimum requirement is that the desired information can be collected under the most challenging conditions so that the data are not biased toward more favorable situations. For example, it might be possible to monitor the behavioral states of three individuals simultaneously when they are resting but not when they are very active. To collect unbiased data, the researcher would have to observe no more than the number of individuals for which it is possible to record the necessary data consistently under all conditions. Usually this means just watching a single animal at a time. Otherwise, the observer risks oversampling specific individuals or undersampling everyone.

Researchers employ a number of different sampling methods to study chimpanzees. In general, different sampling procedures are used for investigating behavioral events versus behavioral states. It is typical, however, to collect event and state data simultaneously. Several of the most common methods are described in the following sections.[2]

Focal-Animal Sampling

In focal-animal sampling, the researcher collects data on the behavior of a single animal, called the "target" or "focal," for a predetermined period of time. Sampling periods can range from minutes to days for each target. During the sample period, the observer restricts his or her attention to the actions and interactions of the target only, whatever else other animals may be doing nearby and no matter how exciting other events may seem. Data may be recorded at predetermined time intervals ("instantaneous" sampling) or continuously, depending on what kinds of information the researcher is interested in. At the end of the sample period, the observer switches to a new target following one of a variety of possible rotation schedules. Examples of target rotation schedules include a completely randomized list of all members of the group, alternating between different age/sex classes with individuals randomized within classes, or targeting animals in the order they first come into view on a given day.

Focal-animal sampling is the most common sampling procedure used in long-term studies of well-habituated chimpanzees. There are two reasons for this. First, by choosing targets and setting sample periods according to predetermined schedules, researchers can avoid inadvertently biasing observations toward particular animals, especially more habituated, active, or otherwise flamboyant individuals. Second, researchers can use focal-animal sampling to obtain several different types of data amenable to statistical analysis and useful for testing hypotheses. For example, investigators can use focal-animal sampling to generate data to determine frequencies and rates of behavioral events, durations of behavioral states, and components of multi-event behavioral sequences.

A common type of data generated by focal-animal sampling is the proportion of time animals devote to their primary behavioral states, or their "activity budget." A basic activity budget might include the proportion of waking time spent feeding, resting, traveling between foraging sites, and socializing. To determine the activity budgets of different classes of individuals within a population over time, a researcher could record the behavior of focal animals every minute on the minute, regardless of what they do in the fifty-nine seconds between sample points. With a large enough sample, an accurate picture of each animal's daily activity budget can be generated. By recording the animal's behavioral state in this way, the researcher is then freed up to record other kinds of data during the time between instantaneous sample points.

Scan Sampling

Instantaneous sampling can be applied either to individuals (instantaneous focal-animal sampling) or to groups (scan sampling). In scan sampling, an observer records the behavior of all individuals in a group simultaneously at predetermined intervals. The numbers and types of behavior that can be monitored reliably in this way depend on the size of groups, observation conditions, and group dynamics. For habituated chimpanzees, scan sampling is effective only for fairly simple types of behavioral data. This is because visibility is often poor in the forest, with individuals in larger groups constantly coming in and out of view during periods of social activity. Consequently, it can take too much time to collect the necessary data on all individuals in a group simultaneously, such that the sample is no longer "instantaneous" but spans many minutes, even running into the next sample period. Scan sampling in chimpanzees therefore typically involves comparatively long intervals, from fifteen to sixty minutes. The data collected most often by scan sampling are the number, identity, and proximity of individuals in subgroups throughout the day.

Sequence Sampling

In sequence sampling, the observer monitors the progression of a specific behavior or behavioral interaction. This allows the researcher to explore patterns of social behavior that would be missed in focal-animal or scan sampling. For example, focal-animal sampling could reveal how many times per hour a female copulated during her fertile period. However, it would be inadequate for investigating the full range of events and individuals involved in mating behavior. By focusing on behavioral sequences rather than just the act of copulation, the observer would be able to collect a variety of data concerning actions preceding and following copulation. This could include male mate-guarding behaviors; male solicitation gestures; aggression by males before, during, or after copulation; female solicitation behaviors; female responses to aggression; female resistance to mating; affiliative behaviors before and after copulation; and so on.

Ad Libitum Sampling

In *ad lib* sampling, the observer records behaviors of interest without following predetermined rules regarding targets or time intervals. This method of information gathering is typical of the early stages of a field study, before the study animals are well habituated or can be easily identified. It is useful for gaining a general picture of behavioral categories and sequences that may subsequently help researchers develop testable hypotheses and identify the kinds of data that will be most reliably obtained as the study progresses. Over the course of a study, *ad lib* observations can also produce an important record of rare behavioral events, providing a glimpse of behavioral potential often not revealed by other sampling techniques.

Because *ad lib* sampling is not systematic, it is not useful for quantitative and comparative analyses of behavior. It therefore plays a minor role in field studies designed to test hypotheses proposed to explain the adaptive significance of behaviors. Nonetheless, chimpanzees are extremely innovative, and *ad lib* observations remain central to the ongoing documentation of their behavioral diversity and complexity, even at sites where they have been watched for many decades.[3]

A Note on Pooling Data

Combining data from all individuals into a single sample is referred to as "pooling" the data. Pooling is unacceptable in most types of analyses because if there are more observations on some individuals than others, their behavior will be overrepresented in the analysis. Statistical tests include calculations that take into account differing numbers of observations made on different individuals. To take advantage of these tests, researchers must therefore collect data on known individuals only and enter their data for testing in a way that keeps observations linked to the individual they are from. It is for this reason above all that researchers must be able to distinguish their study animals individually. A practical implication of this necessity is that new field workers must commit substantial amounts of time in the field learning to identify their animals before they can begin to collect most types of behavioral data.

ADDITIONAL TYPES OF DATA

Scientists studying chimpanzees collect many types of information. The focus of this book is social behavior, and thus each chapter primarily presents observational data concerning interactions between individuals. Nevertheless, to investigate the causes and consequences of individual behavior patterns, other sorts of empirical evidence are often helpful. These include ecological measurements and analyses of physical samples from the animals themselves used to assess their physiological condition and to determine their genetic relationships.

Ecological Measurements

Perhaps the most important ecological data typically collected by field workers are measurements of food abundance, distribution, and quality. Common methods employed to assess food abundance and distribution include regularly monitoring the number and ripeness of fruits on individually marked trees throughout a study area, or on trees located within sample plots evenly distributed in the study area and surveyed on a scheduled basis. Food quality can be evaluated through biochemical analyses of specimens taken from the field and brought to research laboratories in home institutions. When feasible, researchers also seek information about the location and composition of neighboring chimpanzee groups, and about the identity and population characteristics of non-chimpanzee competitor, predator, and prey species. Seasonal variation in rainfall may also be relevant to analyses of social behavior.

Physical Data

Field workers typically collect a variety of physical samples from their study animals for laboratory analysis. Hair is collected from night nests, urine can be collected using plastic sheets held beneath individuals in the canopy, and dung is typically collected from the forest floor in plastic bags. DNA profiles can be generated from hair, urine, and feces to determine genetic relationships among group members and the reproductive success of individuals. Chemical analyses of urine and fecal samples can provide information about physiological condition. For example, elevated levels of the steroid hormone cortisol are associated with stress, the presence of urinary ketones indicates nutritional shortages, and estrogen and progesterone concentrations indicate female reproductive state. Dung is also examined for the presence of gut parasites and is a key source of information about rarely eaten foods.

THE IMPORTANCE OF LONG-TERM DATA COLLECTION

Long-term field sites build up substantial databases that make it possible to explore a variety of questions that would otherwise be impossible to address. Demographic records can be analyzed to determine birth and death rates in relation to environmental variation; the social transmission of behavior patterns can be tracked and compared between populations living in different habitats; the effects of long-term trends in community ecology, including impacts from human population pressure and climate change, can be monitored; and so on. Continuous observation of a population over a long period of time is also necessary to document rare behaviors and improve our understanding of chimpanzee behavioral diversity and the ability of chimpanzees to adapt to ecological change. All in all, long-term field projects have now made it possible to ask the kinds of evolutionary questions that have been routine for decades in studies of species with shorter generation times.

End Notes

PREFACE

1. McGrew (2017: Table 1) lists 120 field sites in twenty countries where chimpanzees have been studied, spanning habitats ranging from rain forest to savanna.
2. For example, Kühl *et al.* (2017) have recently documented an 80 percent decline in the critically endangered western chimpanzee (*Pan troglodytes verus*) population.
3. Pusey *et al.* (2008a), Wrangham and Ross (2008).
4. For Gombe, Goodall (1986); for Mahale, Nakamura *et al.* (2015); for Bossou, Matsuzawa *et al.* (2011); for Taï, Boesch and Boesch-Achermann (2000); for Budongo, Reynolds (2005). There are also several books aimed primarily at specialists that emphasize behavioral diversity between chimpanzee populations: Wrangham *et al.* (1994), McGrew *et al.* (1996), Boesch *et al.* (2002), McGrew (1992, 2004).
5. Bonobos inhabit forests in the Democratic Republic of the Congo and are geographically separated from common chimpanzees by the Congo River. Though similar to common chimpanzees in many ways, they exhibit key behavioral differences that are the focus of ongoing comparative socioecological analyses. See Furuichi and Thompson (2008) and references therein.
6. See reviews in Lonsdorf *et al.* (2010).
7. De Waal (1982), Aureli and de Waal (2000), Aureli *et al.* (2012), Jensen (2012).
8. De Waal (1994), Kummer (1995:132).

1 PRIMATES, APES, AND THE STUDY OF CHIMPANZEE SOCIAL BEHAVIOR

1. According to the IUCN/SSC Primate Specialist Group (2017) listing, 496 species and 695 species and subspecies overall (www.primate-sg.org/species/).
2. Strier (2011), Fleagle (2013).
3. Dominy *et al.* (2003).
4. Following Groves (2001). Note that an alternative division, between the Prosimii and Anthropoidea, is followed by many authors. The difference between the two taxonomies lies with the placement of tarsiers: the prosimians include the strepsirrhines plus tarsiers, and the anthropoid primates are the haplorrhines minus tarsiers.

5. The first primate-like animals appeared during the Paleocene epoch (66 to 56 million years ago, or mya). Large numbers of prosimian-like species evolved during the Eocene (56 to 34 mya), followed by multiple monkey radiations in the Oligocene (34 to 23 mya). See Gradstein (2012) for geologic time scale and research methods, and Fleagle (2013) for primate evolution.

6. Strepsirrhines retain the *rhinarium*, a moist nasal membrane connected to the upper lip that has specialized olfactory receptors. This structure is common in other animals with well-developed olfaction, e.g., dogs and cats. Strepsirrhines also retain the *tapetum lucidum*, a specialized membrane in the eye that enhances night vision and causes "eye shine" when, for example, exposed to a flashlight (or, in the case of a nocturnal mammal traversing a road, a car headlight). Among the strepsirrhines, members of the subfamilies Lorisinae and Perodicticinae have lost the ability to leap (Nekaris and Bearder, 2011). Note that like strepsirrhines, tarsiers are small, nocturnal, and less gregarious, but like the haplorrhines, they lack the *rhinarium* and *tapetum lucidum*.

7. Kappeler (2012: 27).

8. Fleagle (2013).

9. Kaas (2005).

10. The Miocene epoch, sometimes referred to as the "Age of Apes," spanned the period from 23 to 5 mya. Fleagle (2013: Tables 15.1–15.3) lists eighty-two possible fossil ape species during this period. See contributions in Hartwig (2002) for more conservative estimates.

11. The taxonomy of the Hylobatidae is currently uncertain. Groves (2001) recognizes fourteen species; Geissmann (2002) recognizes twelve.

12. Groves (2001).

13. Carbone *et al.* (2014).

14. Hey (2010), Bjork *et al.* (2011), Scally *et al.* (2012), Moorjani *et al.* (2016), Amster and Sella (2016).

15. The taxonomic status of fossil bipedal hominins remains hotly disputed and constantly under revision, both at the generic and species levels. For example, Wood (2010) lists seven genera and twenty-three species, but notes that these can be collapsed to four genera and seven species.

16. Russon and Begun (2004), Smaers and Soligo (2013); but see Tomasello and Call (1997) for a more cautious evaluation of the difference between monkeys and apes in intelligence.

17. Kraft *et al.* (2014).

18. Robbins (2011: Table 19.2).

19. Mahale: Itoh and Nakamura (2015: 234), who review dietary repertoires reported from several field sites.

20. For example, chimpanzees in Kibale National Park: Balcomb *et al.* (2000), Watts *et al.* (2012a).

21. See Chapman *et al.* (1999) and Itoh and Muramatsu (2015) for data from the Kibale and Mahale Mountains National Parks in Uganda and Tanzania, respectively.

22. For average daily travel distances and home range estimates across studies and sites, see the following reviews: for gibbons, Bartlett (2011: Tables 17.2 and 17.3); for orangutans, Singleton *et al.* (2009: Tables 13.1 and 13.3; Utami Atmoko *et al.*, 2009: Table 15.1); for gorillas, Robbins (2011: Table 19.2); for chimpanzees at long-term field sites, Stumpf (2011: Table 20.3).

23. Chimpanzees in at least one population are known to nest regularly on the ground, the reason for which remains unclear (Koops *et al.*, 2007, 2012).

24. Nest building is a comparatively complex task that involves the choice of a suitable location and the manipulation of branches into a sort of inter-locking mesh. Infants watch their mothers and begin practicing from an early age, though it takes several years before they can complete an adequate construction. See Prasetyo *et al.* (2009) for a comparative review of great ape nest building behavior.

25. See Zamma (2014) for night activity in Mahale chimpanzees. Because apes rarely travel at night, researchers can locate study animals easily at daybreak if they had followed them to their nesting site the previous evening (see the Appendix for data collection techniques).

26. The terms and concepts used to describe animal societies have been used in different ways over the years by different researchers. The schema adopted here follows Kappeler and van Schaik (2002), who review previous theoretical contributions and attempt to standardize the terminology.

27. Gibbon groups were originally called "families" but more recently have been referred to as "households" since adult membership can change through death or desertion and the young may not be related to both adults. Reviewed in Bartlett (2011).

28. Mitra Setia *et al.* (2009).

29. Harcourt and Stewart (2007: 135).

30. Reichard and Barelli (2008).

31. Harcourt and Stewart (2007: 132–134, 187)

32. It is typical in group-living primates for members of the sex that does not disperse to form the strongest social relationships, as occurs in chimpanzees. The reason why bonobos depart from this pattern remains unclear. See, for example, Parish (1996) and White and Wood (2007).

33. See examples in Biro *et al.* (2010b) and Hobaiter *et al.* (2014a).

34. For example, see Goodall (1986) for detailed descriptions of infant and juvenile development in chimpanzees, and Furuichi (1989) for analysis of the importance of maternal support for adult sons competing for dominance status in bonobos.

35. For example, elephants, hyenas, and marine mammals. See contributions in de Waal and Tyack (2003).

36. Reviewed in Seyfarth and Cheney (2003).

37. Watts *et al.* (2000).

38. Kummer (1968) described "tripartite relations" in hamadryas baboons (*Papio hamadryas*). de Waal (1982: 175) described triadic awareness as "the capacity to perceive social relationships between others so as to form varied triangular relationships."

39. Crockford *et al.* (2007).

40. See Aureli *et al.* (2008) for a general review of fission–fusion dynamics and Connor (2007) for dolphins in particular. See Wittig *et al.* (2014b) for experimental evidence of triadic awareness in wild chimpanzees (Budongo).

41. The idea that larger brains evolved because of the advantages of social sophistication is known as the "Social Brain Hypothesis": see Humphrey (1976), Dunbar (1998), Dunbar and Shultz (2007), Cheney and Seyfarth (2007), and Grueter (2015). A competing idea is that larger brains were advantageous for finding high-quality foods, which then provided the necessary energy to support expensive brain tissue (Milton, 1993). The two hypotheses are not necessarily mutually exclusive.

42. Hinde (1976).

43. Steklis and Kling (1985: 94) define affiliative social interactions as "those behaviors that promote the development of and that serve to maintain

social bonds." Agonistic behaviors include a wide range of behaviors related to fighting, including physical and vocal threats, attacks, defensive behaviors, escape, submissive signaling, and so on (Scott and Fredericson, 1951).
44. See Silk (2012: 556) for a schematic representation of primate social relationships and a discussion of the difficulty of comparing primate social bonds to human friendships.
45. For example, researchers continue to document new behaviors among chimpanzees despite decades of research by a multitude of observers at many field sites. See Nishida *et al.* (2010) for a recent ethogram of chimpanzees which includes more than 1,000 behaviors observed at study sites across Africa.
46. By comparison, members of permanent social groups in most primate species stay in visual or auditory contact with each other throughout the day.
47. As described in detail in Chapter 4, for example, adult male chimpanzees generally spend more time around other males than they do around females or than adult females spend around other females.
48. See Goodall (1968a: 201–202, 263–269; 1986: 387–408) for detailed descriptions of methods of initiation and the precise movements involved in social grooming at Gombe, along with examples of individual variation in patterns of participation. Chimpanzees also groom themselves, which is referred to as "autogrooming" or "self-grooming," often while they are being groomed by another individual.
49. Gombe: Goodall (1986: 388 and Fig. 14.3); Taï N: Boesch and Boesch-Achermann (2000: Figs. 6.5 and 6.7).
50. Dunbar (1988), Lehmann *et al.* (2007).
51. Dunbar (1988).
52. Goosen (1987), Schino *et al.* (1988), Aureli *et al.* (1999), Crockford *et al.* (2008), Wittig *et al.* (2008).
53. Dunbar (1988: 254), Henzi and Barrett (1999).
54. Budongo: Crockford *et al.* (2013).
55. See Goodall (1986: 369–372) for detailed descriptions from Gombe.
56. Mahale: see Hayaki (1985) and Matsusaka (2004) for two quantitative analyses of play, and Matsusaka *et al.* (2015) for a review; Budongo: see Fröhlich *et al.* (2017) for a study of the influence of play on gesture development.
57. Ngogo: Watts and Mitani (2001).
58. Ngogo: Watts (2002), Mitani (2006b).
59. See Goodall (1968a: 275–281; 1986: Chapters 12 and 13) for detailed descriptions of fighting, threat, and submissive behaviors in Gombe chimpanzees. See Nishida *et al.* (2010) for an exhaustive listing of agonistic behaviors.
60. See Goodall (1968a: 281–284, 1986: 358–367) for detailed descriptions of contact and reassurance behaviors in Gombe chimpanzees. See Nishida *et al.* (2010) for a more extensive listing.
61. Gombe: Goodall (1971: 118; 1986: 361–364).
62. Initial studies by de Waal and van Roosmalen (1979) on chimpanzees and by de Waal and Yoshihara (1983) on rhesus macaques (*Macaca mulatta*) spawned a new area of research on primate post-conflict interactions generally. The basic post-conflict/matched-control (PC/MC) methodology developed in these studies is still employed today, with various modifications. The basic principle is that the behaviors of fight participants are observed for a set period of time following the aggressive interaction and compared to their behaviors on a subsequent day when no fight has taken place. Reconciliation and consolation are deemed to have occurred if friendly contact occurs post-conflict sooner than during the matched control. See Arnold *et al.* (2011) and

Aureli *et al.* (2012) for reviews of the extensive literature on reconciliation and consolation, now documented in more than thirty primate species.

63. Strict match-controlled procedures are difficult to implement in the field because the locations and associations of individuals are unpredictable from day to day. Studies of reconciliation and consolation in the wild more often simply rely on the observation of peaceful post-conflict contact within a short time interval, usually ten minutes, after an aggressive incident. Gombe: Goodall (1986: Table 13.1, no time interval reported); Mahale: Kutsukake and Castles (2004, ten-minute interval); Taï N: Wittig and Boesch (2005, relative to average time between affiliative interactions); Kanyawara: Muller (2002, ten-minute interval); Budongo: Arnold and Whiten (2001, PC/MC, but the day of MC variable); Ngogo: Watts (2006, within five minutes).

64. Numerous studies in primates and other mammal species have demonstrated that peaceful post-conflict contact reduces stress (evidenced by lower rates of self-directed behaviors such as self-scratching, lower cortisol levels, and lower heart rates), increases the likelihood of resource sharing, and lowers the likelihood of renewed aggression. See reviews in Aureli and Smucny (2000), Aureli *et al.* (2002), and Aureli *et al.* (2012).

65. The idea that social relationships in group-living primates are a resource worth investing in, with benefits that outweigh their costs, was first formally advanced by Kummer (1978). See Silk (2012) for a review of the now extensive literature on the reproductive benefit of social relationships.

66. Ngogo: Watts (2006).

67. Note that two types of third-party contact can be distinguished. "Solicited consolation" refers to approach by a fight participant to a third individual. Since such approaches sometimes involve efforts to recruit allies, and thus include a confounding contextual variable, they are not included in functional analyses of consolation. Such analyses therefore are restricted to approaches by a third party to a fight participant (de Waal and Aureli, 1996).

68. Gombe: Goodall (1968a: 283); Chester Zoo (Chester, UK): Fraser and Aureli (2008), Fraser *et al.* (2008).

69. Koski and Sterck (2007).

70. Burgers' Zoo (Arnhem, the Netherlands): Koski and Sterck (2009).

71. Wittig (2010), Fraser *et al.* (2009); also called "triadic reconciliation" by Judge (1991).

2   SEVEN LONG-TERM FIELD STUDIES

1. Butynski (2003).
2. See Pusey *et al.* (2007, 2008a) and Wilson (2012) for descriptions of the study site and animals. The Kalande community, unfortunately, has recently suffered a precipitous decline (M. Muller, personal communication). Note that the size of a chimpanzee group varies over time as a consequence of births, deaths, immigrations, and emigrations. The area utilized by a group also varies constantly over time in response to demographic and ecological variation as well as competition with neighboring groups. Altitudes range from 766 meters above sea level (m.a.s.l.) at the lakeshore to peaks of 1,300–1,600 m.a.s.l.
3. Pusey *et al.* (2005).
4. Mjungu (2010).
5. Goodall (1986: 58).

6. Pusey *et al.* (2008a).
7. Although both Goodall (1965) and Reynolds and Reynolds (1965) had previously described the temporary nature of chimpanzee subgroups, Nishida (1968) was the first to describe the stable nature of the larger social group of which such subgroups form a part. He called this stable group a "unit-group," but the term "community," used by Goodall (1971), is now more commonly employed.
8. Nishida and Kawanaka (1972).
9. See Nakamura *et al.* (2015) for a massive review of studies at Mahale.
10. Nishida (2012: 15).
11. Nishida (1968: 202).
12. Nakamura *et al.* (2013).
13. Nishida (1990: 21).
14. Nishida and Hiraiwa-Hasegawa (1985).
15. Baldwin and Teleki (1973).
16. Sugiyama and Fujita (2011).
17. Sakuru and Matsuzawa (1991).
18. Humle (2011).
19. Hockings *et al.* (2010).
20. Boesch and Boesch-Achermann (2000).
21. Marchesi *et al.* (1995).
22. Boesch and Boesch (1983).
23. Boesch and Boesch-Achermann (2000: 25).
24. Boesch (2008).
25. Boesch (2008).
26. Boesch *et al.* (2008).
27. Köndgen *et al.* (2008).
28. Personal communication from R. Wittig, director of the Taï Chimpanzee Project, June 2013.
29. Boesch (2008).
30. Isabirye-Basuta (1989).
31. Ghiglieri (1984).
32. Wrangham *et al.* (1992).
33. Hyeroba *et al.* (2011).
34. Struhsaker (1975).
35. Wilson *et al.* (2012), Muller and Wrangham (2014).
36. Wrangham (2008).
37. Hyeroba *et al.* (2011).
38. Reynolds (2005).
39. Plumptre *et al.* (2010).
40. Fedurek *et al.* (2013a), Hobaiter *et al.* (2014b).
41. Reynolds (2005: 164, 170).
42. Wrangham (2008).
43. Ghiglieri (1984).
44. Watts (2012).
45. Mitani *et al.* (2010).
46. Watts (2015).
47. Watts *et al.* (2012a, b), Wood *et al.* (2017).
48. See McGrew (2017:Table 1) for a list of 120 field sites in twenty countries.
49. Fongoli: Preutz *et al.* (2015); Semliki: McGrew *et al.* (2007); Gashaka Gumti: Fowler and Sommer (2007); Bulindi: McLennan (2011).

3   CHIMPANZEE FISSION–FUSION SOCIAL ORGANIZATION AND ITS CONSERVATION IMPLICATIONS

1. Nishida (1968), studying at Mahale, was the first to describe the structure of chimpanzee social groups and called them "unit-groups." Goodall (1965: 453) used the term "community" to refer to her study group at Gombe, although she identified mother–offspring associations as the only groups that might be stable over a span of years (Goodall, 1968a: 167, 211). The term "community" is used more commonly today by Western researchers.
2. Stumpf (2011: Table 20.5).
3. One early report from Taï N (Boesch, 1991a) suggested that larger subgroups might be a response to the threat of leopard predation. However, this study did not quantify fruit availability or consumption, important confounding variables, and therefore remains speculative. In any case, predation is not a consistent source of mortality for chimpanzees at any field site (see Jenny and Zuberbühler, 2005; Nakazawa et al., 2013).
4. Pusey (1980); Pusey and Packer (1987); Pusey and Wolf (1996).
5. Occasional male migration is suspected in the very small and isolated population at Bossou, where there are no neighboring groups to restrict male movement (Sugiyama, 1999).
6. Emery Thompson (2013: Table I); Bossou females mature somewhat earlier, with first swelling occurring between the ages of eight and nine years (Sugiyama, 2004: Table 1).
7. At Taï N and Budongo, a single natal female at each site has so far been reported to stay and breed within their community (Boesch and Boesch-Achermann, 2000: 48; Reynolds, 2005: 102). At Mahale, two natal females were suspected to have bred in M-group early in the study (Hiraiwa-Hasegawa et al., 1984: 410). Subsequently, five females did not emigrate during a brief period during which an epidemic reduced the community size by half (Nishida, 2012: 192). At Kanyawara, three natal females have stayed to breed, two of whom were daughters of alpha females (Kahlenberg et al., 2008b; R. Wrangham, personal communication 8/29/17). As of 2013 at Ngogo, three natal females had bred within the community (Langergraber et al., 2009: 846; Langergraber, personal communication 9/11/13); Wood et al. (2017: 44) report that "many natal females have remained at Ngogo as adults and reproduced there."
8. Knott (2001), Atsalis and Videan (2009).
9. Kanyawara: Emery Thompson and Wrangham (2008); Emery Thompson et al. (2012).
10. Gombe: Goodall (1986: 342, Table 12.3), Miller et al. (2014); Mahale: Nishida (1989: 81); Taï N: Wittig and Boesch (2003); Kanyawara: Muller (2002).
11. Gombe: Pusey et al. (1997), Williams et al. (2002a), Murray et al. (2006); Mahale: Nishida (1989: Table 8); Taï N: Wittig and Boesch (2003); Taï NS: Riedel et al. (2011); Kanyawara: Emery Thompson et al. (2007a), Kahlenberg et al. (2008a).
12. Kanyawara: Muller (2002: Fig. 8.1), with comparisons to Gombe and Mahale (Tables 8.1 and 8.2).
13. Gombe: Goodall (1986); Mahale: Nishida (1989); Taï N: Boesch and Boesch-Achermann (2000: 45); Kanyawara: Muller (2002: 119), Kahlenberg et al. (2008b); Budongo: Townsend et al. (2007).
14. Gombe: Goodall (1986), Pusey et al. (2008b); Mahale: Nishida (1989); Budongo: Townsend et al. (2007).
15. Kanyawara: Kahlenberg et al. (2008a, b).

16. Gombe: Williams (2002a), Murray *et al.* (2007a); Mahale: Nishida (1989).
17. It is usually impossible to know how old immigrant females are when they produce their first offspring because their ages are only estimated. Occasionally, though, when researchers can monitor neighboring communities, known females immigrate into study populations and reproduce. Based on these cases it is possible to compare age at first birth between natal and immigrant females. In nearly all cases, natal females begin to reproduce at earlier ages, implying a short-term reproductive advantage. Gombe: unpublished data cited in Pusey and Schroepfer-Walker (2013: 6); Mahale: Nishida *et al.* (2003) found that the average age at first birth was thirteen years for natal females and fourteen years for immigrants.
18. Gombe: See Wrangham (1974: 91) for estimates of increased food availability from provisioning. Note that even though female non-dispersal rates have persisted since the termination of provisioning in 2000, this is not evidence that provisioning had no effect, since a new behavior pattern could have been established, and habitat reduction is an ongoing constraint (see also Chapter 11 for the establishment of new and enduring behavior patterns).
19. Mahale: Nishida (2012: 192).
20. Kanyawara: Kahlenberg *et al.* (2008a: 943).
21. Ngogo: See Watts *et al.* (2012a) for habitat richness.
22. Pusey (1980, 1990); Charlesworth and Willis (2009).
23. "Estrus" refers to the period around the time of ovulation in a female's menstrual cycle. The most striking change in female appearance during estrus is a dramatic swelling of the anogenital area. These "sexual swellings" begin several days before ovulation, peak in size around the time of ovulation, and then decrease for several days following ovulation. Thus, they provide an important indicator of a female's reproductive state, are highly attractive to males, and are easily monitored by researchers.
24. Gombe: Goodall (1986: 466).
25. Mahale: Nishida (1990: 32).
26. Ngogo: Langergraber (2013; personal communication)
27. Gombe: Goodall (1986: 469).
28. Kanyawara: Stumpf *et al.* (2009).
29. Mother–son matings at Gombe: Goodall (1986: 467); at Bossou: Sugiyama and Koman (1979: 336), Nakamura (2011: 258). Goodall (1986: 269) reported several presumed father–daughter copulations at Gombe.
30. Constable *et al.* (2001) employed noninvasive paternity testing using hair and fecal analysis to confirm the relationship between the father and daughter that produced an offspring at Gombe.
31. Bossou: Sugiyama and Fujita (2011).
32. Kanyawara: Stumpf *et al.* (2009).
33. Shimada *et al.* (2009).

4   SEX DIFFERENCES IN RANGING AND ASSOCIATION PATTERNS

1. Taï S: Janmaat *et al.* (2013, 2014) and Ban *et al.* (2014, 2016) provide evidence that Taï chimpanzees monitor the fruiting status of trees throughout their range, periodically checking specific, previously utilized trees for the presence of ripe fruits and adjusting their foraging routes accordingly.

2. Long-term records of the monthly fruiting status of hundreds of individual trees now exist at most of the long-term field sites, permitting analyses of the relationship between fruit availability and party size. Gombe: Wrangham (1977), Murray *et al.* (2006); Mahale: Matsumoto-Oda *et al.* (1998); Taï N: Doran (1997), Anderson *et al.* (2002, 2005), Polansky and Boesch (2013); Taï S: Wittiger and Boesch (2013); Kanyawara: Chapman *et al.* (1994); Budongo: Newton-Fisher *et al.* (2000); Ngogo: Mitani *et al.* (2002a), Potts *et al.* (2011).
3. Stumpf (2011: Table 20.3).
4. Gombe: Wrangham (1979), Williams *et al.* (2002a); Mahale: Hasegawa (1990); Taï N: Lehmann and Boesch (2005); Kanyawara: Chapman and Wrangham (1993); Budongo: Fawcett (2000) in Reynolds (2005: 103).
5. For example, about one patrol/month on average at Gombe (Goodall, 1986: Table 17.1) and Taï N (Boesch and Boesch-Achermann, 2000: Table 7.11) and about three patrols/month at Ngogo (Watts and Mitani, 2001).
6. Gombe: Williams *et al.* (2002b); Mahale: Hunt (1989); Taï N: Doran (1997); Kanyawara: Pontzer and Wrangham (2006).
7. Wrangham (2000).
8. Gombe: Wrangham (1979), Williams *et al.* (2002a), Murray *et al.* (2006); Mahale: Nishida (1989), Hasegawa (1990); Kanyawara: Emery Thompson *et al.* (2007a); Budongo: Reynolds (2005: 103); Ngogo: Langergraber *et al.* (2009).
9. Gombe: Murray *et al.* (2008).
10. Gombe: Goodall (1986); Mahale: Nishida (1979); Taï N: Doran (1997), Lehmann and Boesch (2008); Kanyawara and Budongo: Emery Thompson and Wrangham (2006); Ngogo: Mitani *et al.* (2002a).
11. Gombe: Wrangham and Smuts (1980).
12. Gombe: Williams *et al.* (2002b).
13. Gombe: Murray *et al.* (2007a).
14. Taï N: Boesch (1996), Lehmann and Boesch (2004); Taï S: Wittiger and Boesch (2013).
15. Following Nishida's (1968: 194) "index of familiarity," the dyadic association index, or DAI, for individuals A and B is calculated as follows: DAI = $(c \mid a + b + c)$, where a is the number of sample points in which A is observed without B, b is the number of sample points in which B is observed without A, and c is the number of sample points in which A is observed in association with B.
16. Gombe: Wrangham and Smuts (1980), Goodall (1986: Fig. 7.3); Mahale: Nishida (1968: 197); Taï N: Lehmann and Boesch (2008); Kanyawara: Wrangham *et al.* (1992), Gilby and Wrangham (2008); Budongo: Newton-Fisher (2003); Ngogo: Pepper *et al.* (1999).
17. Taï N: Lehman and Boesch (2008); Ngogo: Langergraber *et al.* (2009).
18. Gombe: Pusey (1980); Kanyawara: Stumpf *et al.* (2009).
19. Gombe: Pusey (1980); Mahale: Nishida (1979), Nishida *et al.* (1990); Taï N: Boesch and Boesch-Achermann (2000: 46); Kanyawara: Stumpf *et al.* (2009) suggest the likelihood of visiting in view of prolonged absences of natal adolescent females; Budongo: Reynolds (2005: 31).
20. Gombe: Pusey (1990: 213).
21. Gombe: Pusey (1980, 1990: 242), Goodall (1986), Pusey *et al.* (2008b); Mahale: Nishida (1989); Taï N: Boesch and Boesch-Achermann (2000: 45); Kanyawara: Muller (2002: 119), Kahlenberg *et al.* (2008b); Budongo: Reynolds (2005: 119, 132).
22. Wallis (1992, 1997).
23. Reviewed in Emery Thompson (2013).

24. Gombe: Wrangham (1977), Goodall (1986: Figs. 9.3, 7.2); Mahale: Hasegawa (1990), Matsumoto-Oda et al. (1998); Taï S: Wittiger and Boesch (2013); Kanyawara: Machanda et al. (2013); Budongo: Reynolds (2005); Ngogo: Mitani et al. (2002a).

25. Gombe: Tutin (1979), Goodall (1986); Mahale: Hasegawa and Hiraiwa-Hasegawa (1983, 1990); Taï N: Boesch and Boesch-Achermann (2000); Kanyawara: Wrangham (2002); Budongo: Reynolds (2005: 119); Ngogo: Watts (2015: 166).

26. Gombe: Murray et al. (2009).

27. Gombe: Williams et al. (2002a: 357–358), Murray et al. (2006); Mahale: Nishida (1989); Kanyawara: Emery Thompson et al. (2007a), Gilby and Wrangham (2008), Kahlenberg (2008a), Machanda et al. (2013); Ngogo: Wakefield (2008, 2013), Langergraber et al. (2009).

28. Taï NS: Riedel et al. (2011), Wittiger and Boesch (2013).

29. Gombe: Murray et al. (2006, 2007a); Mahale: Nishida (1989); Kanyawara: Emery Thompson et al. (2007a); but see Reynolds (2005: 104) for a different pattern at Budongo.

30. Reviewed in Wrangham (2000).

31. Bossou: Sugiyama and Fujita (2011).

32. Bossou: Sugiyama (1988).

33. Taï N: Compare Doran (1997) and Lehmann and Boesch (2004).

34. Hockings (2011) suggests that crop raiding permits group cohesiveness at Bossou. At Kanyawara, Emery Thompson et al. (2014) reported that females suffered significant feeding costs when many males were present in parties, which could explain why females at Taï became more cohesive when fewer males were present in the community.

35. Taï N: Lehmann and Boesch (2004).

36. Kanyawara: Otali and Gilchrist (2006).

37. Ngogo: Mitani and Amsler (2003), Mitani (2006a).

38. Potts et al. (2011), Watts et al. (2012b).

39. Langergraber et al. (2013: 870).

5   FEMALE SOCIAL RELATIONSHIPS

1. In a schema followed by many researchers, Goodall (1986: Table 5.1) and Pusey (1990) categorized the developmental stages of chimpanzees as follows: (1) *Infancy* (zero to five years): birth to weaning, characterized by initially constant carrying by the mother followed by progressive locomotor competence; (2) *juvenile* (five to eight years): weaning until onset of testicular growth in males and first anogenital swelling in females, still in close proximity to mother all of the time, including occasional riding on mother during travel; (3) *early adolescence* (eight to twelve years): completion of testicular growth in males and first menstrual bleeding in females, progressively more independent ranging; (4) *late adolescence* (eleven to fifteen years): attainment of full body size in males, from first estrus cycle to first conception in females; (5) *adulthood* (fifteen to thirty-five years); (6) *old age* (thirty-five to forty-plus years).

2. Bossou: Matsuzawa (2011) observed a juvenile female carry and pat a 50 × 10 cm stick while following her mother who was carrying a sick infant; Kanyawara: Kahlenberg and Wrangham (2010).

3. Gombe: Lonsdorf et al. (2014a, b).

4. Pusey (1990: 236).

5. Gombe: Lonsdorf et al. (2004), Lonsdorf (2005).

6. Gombe: Pusey (1983: 371); Mahale: Hiraiwa-Hasegawa (1989).
7. Gombe: Goodall (1986: Figure 10.6, Tables 10.4–10.7), McGrew (1979); Mahale: Uehara (1986), Hiraiwa-Hasegawa (1989).
8. Bossou: Humle *et al.* (2009a).
9. Adolescent males do not establish dominance ranks among themselves, as evidenced by the fact that they rarely utter submissive "pant grunt" vocalizations to one another: at Gombe, Bygott (1979: 414); at Mahale, Kawanaka (1989); at Ngogo, Sandel *et al.* (2017). An earlier study by Sherrow (2012) suggested that adolescent males at Ngogo formed their own dominance hierarchy, but this finding was apparently an artifact of including young adult males in the sample.
10. Gombe: Goodall (1986: 345), Pusey (1990: 230); Mahale: Hayaki *et al.* (1989), Kawanaka (1989); Kanyawara: Muller (2002: 116), Muller *et al.* (2009: Fig. 8.3); Budongo: LaPorte and Zuberbühler (2010).
11. Gombe: Goodall (1986: Table 14.3, Fig. 14.3); Mahale: Nishida (1989: Fig. 2); Taï N: Lehmann and Boesch (2008) for male–male and male–female; Kanyawara: Wrangham *et al.* (1992), Machanda *et al.* (2013); Budongo: Arnold and Whiten (2003) for male–male and male–female; Ngogo: Watts (2000a) for male–male and male–female.
12. Mahale: Nishida (1979: 88).
13. Gombe: Goodall (1986: 182, 399); Mahale: Nishida (1979: 106).
14. Gombe: Goodall (1986: 181); Mahale: Huffman (1990), Takahata (1990a); Taï N: Boesch and Boesch-Achermann (2000: 73).
15. Goodall (1986: 394, 400, Table 14.3).
16. Goodall (1986: 402, Fig. 14.4).
17. Gombe: Goodall (1986: 501), Pusey (1990: 242), Pusey *et al.* (2008b); Mahale: Hayaki (1988: 153), Nishida (1989); Taï N: Boesch and Boesch-Achermann (2000: 45); Kanyawara: Kahlenberg (2008b); Budongo: Reynolds (2005: 119).
18. Gombe: Tutin (1979: 34), Goodall (1986: 398); Mahale: Takahata (1990a); Kanyawara: Wrangham *et al.* (1992), Machanda *et al.* (2013); Budongo: Slater *et al.* (2008); Ngogo: Watts (2000a).
19. Gombe: Tutin (1979: 33), Goodall (1986: 402, 453).
20. Gombe: Goodall (1968a: 211, 1986: 154), Halperin (1979); Mahale: Nishida (1968: 186); Bossou: Sugiyama and Koman (1979); Taï N: Boesch and Boesch-Achermann (2000: 95); Kanyawara: Otali and Gilchrist (2006); Budongo: Reynolds and Reynolds (1965: 399), Reynolds (2005: 89); Ngogo: Ghiglieri (1984: 130), Wakefield (2008).
21. For example, Goodall (1986: 196).
22. Pusey (1990).
23. Williams *et al.* (2002b).
24. Gombe: Goodall (1986: 196), Foerster *et al.* (2015); Mahale: Huffman (1990: 251); Bossou: Sugiyama (1988); Taï N: Lehmann and Boesch (2008, 2009); Budongo: Reynolds (2005: 113); Ngogo: Wakefield (2013), Langergraber *et al.* (2009).
25. Gombe: Goodall (1986: 501), Pusey *et al.* (2008b); Mahale: Nishida (1989: 83); Taï N: Boesch and Boesch-Achermann (2000: 105); Kanyawara: Kahlenberg (2008a); Budongo: Townsend *et al.* (2007).
26. Mahale: Nishida (1989: 83); Taï N: Boesch and Boesch-Achermann (2000: 105, 152), Boesch *et al.* (2008); Budongo: Newton-Fisher (2006).
27. Gombe: Goodall (1986: 358); Kanyawara: Female coalitions against male attackers typically occur when the aggressor is an adolescent male rather than a full adult (M. Muller, personal communication).

28. Gombe: McGrew (1979); Mahale: Nishida (1970: 64); Taï N: Boesch and Boesch-Achermann (2000: 105).
29. Gombe: Bygott (1979), Goodall (1986: 342); Mahale: Nishida (1970); Taï N: Wittig and Boesch (2005: Table 2); Kanyawara: Muller (2002); Budongo: Sugiyama (1969: Table 3), Arnold and Whiten (2001: Table 1).
30. Gombe: Clark (1977: 248).
31. Gombe: Pusey (1983: 370), Goodall (1986: 245); Mahale: Nishida (1970: Tables 10, 11), Nishida and Turner (1996).
32. Gombe: Goodall (1968a: 235); Mahale: Hayaki (1985: 355).
33. Gombe: Bygott (1979: 413), Goodall (1986: 353, Table 12.4), Pusey (1990:230); Mahale: Takahata (1990a); Bossou: Sugiyama and Koman (1979: 333, inferred); Taï NS: Stumpf and Boesch (2006: 761); Kanyawara: Wrangham *et al.* (1992), Muller (2002); Budongo: LaPorte and Zuberbühler (2010).
34. Gombe: Goodall (1986: Fig. 12.4); Mahale: Sakamaki *et al.* (2001); Taï N: Boesch and Boesch-Achermann (2000: 152); Budongo: Newton-Fisher (2006).
35. Taï N: Boesch and Boesch-Achermann (2000: Tables 6.12 and 6.13).
36. Mahale: Nishida and Hosaka (1996: 130).
37. Average interaction rates in seven studies at Gombe, Mahale, Kanyawara, and Ngogo ranged from 0 to 4.6 conflicts per 100 hours of observation: Murray (2007: 858 and Table I); Wakefield (2008) for Ngogo. Conflict rates at Taï are likely to be similar, since interaction rates while feeding were comparable between Taï N and Mahale in two studies (37 and 32 per 100 hours, respectively): Wittig and Boesch (2003: Table III). Rates of pant grunting, a submissive vocalization, were similarly low between females, ranging from 0.01 to 0.02 calls per hour (Murray, 2007: Table I).
38. Mahale: Nishida (1979: Table 8; 1989: 77); Taï N: Wittig and Boesch (2003).
39. Gombe: Goodall (1986: 439), Pusey *et al.* (1997), Murray (2007); Kanyawara: Wrangham *et al.* (1992).
40. Kanyawara: Muller (2002); Budongo: Reynolds (2005: 131); Ngogo: Wakefield (2013).
41. Personal communication from M. Muller, who suggests that the failure to detect these interactions in earlier studies was a consequence of incomplete habituation.
42. Gombe: Markham *et al.* (2014); Kanyawara: Kahlenberg *et al.* (2008b).
43. Gombe: Pusey *et al.* (1997), Jones *et al.* (2010); Kanyawara: Emery Thompson *et al.* (2007a).
44. Gombe: Pusey *et al.* (2008b); Taï NMSE: Boesch *et al.* (2008); Budongo: Townsend *et al.* (2007). Infanticide occurred in the Budongo case.
45. Gombe: Goodall (1986: Table 12.6), Pusey *et al.* (2008b); Taï N: Boesch and Boesch-Achermann (2000: 33), although they were not absolutely certain of the identity of the dead infant; Budongo: Townsend *et al.* (2007). Infanticide was reported in all these studies.
46. Gombe: Goodall (1986: Table 17.2); Mahale: Nishida and Hiraiwa-Hasegawa (1985), Hamai *et al.* (1992); Taï N: Boesch and Boesch-Achermann (2000: 152). Infanticide was reported in the Gombe and Mahale studies.
47. Kanyawara: Arcadi and Wrangham (1999), with infanticide.
48. Kanyawara: Arcadi and Wrangham (1999).
49. Smuts (1986).

6   MALE SOCIAL RELATIONSHIPS

1. Gombe: Pusey (1990: 216).

2. Gombe: Goodall (1986: 368); Mahale: Nishida (1979: 105), Hayaki (1985); Taï NMS: Lehmann *et al.* (2006).

3. Taï NSE: Boesch *et al.* (2010).

4. Gombe: Goodall (1986: Figs. 14.1, 14.3, 14.5); Mahale: Kawanaka (1989), Takahata (1990a); Bossou: Sugiyama (1988: Table III); Taï N: Boesch and Boesch-Achermann (2000: 114, Fig. 6.6); Kanyawara: Wrangham *et al.* (1992: Table 3), Machanda *et al.* (2013); Budongo: Arnold and Whiten (2003), Reynolds (2005: Table 6); Ngogo: Watts (2000a).

5. Gombe: Goodall (1986: 398) recorded an eighty-minute bout between two males; Mahale: Nishida (1970: 57) recorded a 105-minute bout between three males.

6. Gombe: Goodall (1986: 358).

7. Burgers' Zoo (Arnhem, the Netherlands): Koski and Sterck (2007).

8. Gombe: Goodall (1986: 370); Mahale: Takahata (1990b: 162).

9. Ngogo: Mitani (2009a), based on proportion of time spent within 5 m of one another.

10. Gombe: Simpson (1973); Mahale: Takahata (1990b); Taï N: Boesch and Boesch-Achermann (2000:116); Taï S: Gomes *et al.* (2009); Kanyawara: Machanda *et al.* (2014); Budongo: Newton-Fisher and Lee (2011); Ngogo: Watts (2000a).

11. Gombe: Foster *et al.* (2009); Mahale: Takahata (1990b), Nishida and Hosaka (1996: 131); Taï N: Boesch and Boesch-Achermann (2000:124); Kanyawara: Wrangham *et al.* (1992); Budongo: Newton-Fisher (2002); Ngogo: Watts (2000b).

12. Taï S: Gomes and Boesch (2011); Ngogo: Watts (2002), Mitani (2006b).

13. At Gombe, Gilby *et al.* (2013b) found that males who engaged in more coalitionary aggression sired more offspring.

14. Gombe: Simpson (1973), Goodall (1986); Mahale: Kawanaka (1989), Takahata (1990b), but see Kaburu and Newton-Fisher (2013, 2015a) for lack of a grooming bias toward dominants during a period of social instability following the death of the alpha male; Budongo: Arnold and Whiten (2003), Newton-Fisher (2002), Newton-Fisher and Lee (2011); Kaburu and Newton-Fisher (2015a); Ngogo: Watts (2000b).

15. Gombe: Simpson (1973); Mahale: Nishida and Hosaka (1996).

16. Gombe: Foster *et al.* (2009).

17. Kin selection is a form of natural selection that explains the evolution of behaviors in which individuals incur costs while helping relatives. An individual's "inclusive fitness" refers to his or her contribution to the population gene pool, through their own offspring and/or through the offspring of close relatives, since relatives share genes identical by common descent (Hamilton, 1964).

18. Goodall (1986: Table 4.1).

19. Gombe: Goodall (1986: 178) describes the close relationship between two maternal brothers at Gombe separated by six years. With the support of his brother in dominance contests, the younger of the two rose to the highest-ranking position in the community, which he maintained until the death of his older brother; Ngogo: Langergraber *et al.* (2007), Mitani (2009a).

20. Kanyawara: Goldberg and Wrangham (1997); Ngogo: Mitani *et al.* (2000a, 2002b).

21. Gombe: Simpson (1973) for rank; Mahale: Kawanaka (1989), Nishida and Hosaka (1996) for age; Ngogo: Watts (2000b), Mitani (2009a).

22. Taï N: Boesch and Boesch-Achermann (2000: 119); Kanyawara: Wrangham *et al.* (1992); Budongo: Arnold and Whiten (2003); Ngogo: Watts (2000a).

23. Gombe: Goodall (1986: 371); Mahale: Takahata (1990a: 139).
24. Taï N: Boesch and Boesch-Achermann (2000: Table 6.12).
25. Kanyawara: Muller (2002: Fig. 8.1).
26. Gombe: Goodall (1986: 341).
27. Gombe: Bygott (1974, in Goodall, 1986: 416); Mahale: Kawanaka (1989), Nishida and Hosaka (1996), although Takahata (1990b) found no relationship between rank and display rate; Taï N: Boesch and Boesch-Achermann (2000: 112); Kanyawara: Muller (2002), Muller and Wrangham (2004a); Budongo: Reynolds (2005: 126).
28. Gombe: Bygott (1979), Goodall (1968a: 215, 1986: 332); Mahale: Nishida (1968: 191); Kanyawara: Muller (2002); Budongo: Reynolds and Reynolds (1965: 413).
29. Gombe: Bygott (1979), Goodall (1986: Table 12.3); Kanyawara: Muller (2002: 121, Table 8.2).
30. M. Muller, personal communication.
31. Gombe: Goodall (1986: Table 12.4); Kanyawara: Muller (2002: Table 8.3).
32. Kanyawara: Otali and Gilchrist (2006).
33. Gombe: Goodall (1986: Table 12.4); Mahale: Nishida (1970: Table 1b), Takahata (1990b: Table 7.2a); Taï N: Boesch and Boesch-Achermann (2000: Table 6.3); Kanyawara: Muller (2002: Table 8.3).
34. For example, see Goodall's (1986) detailed descriptions of contests and injuries at Gombe.
35. Goodall (1986: 506–510) describes in some detail several prolonged, lethal gang attacks at Gombe.
36. Gombe: Goodall (1986: Fig. 15.1); Taï N: Boesch and Boesch-Achermann (2000: Tables 41. and 6.3); Budongo: Arnold and Whiten (2003); Ngogo: Watts (2000b), Mitani *et al.* (2002b). Note that because of the large number of males at Ngogo, researchers are often unable to distinguish the relative positions of some closely ranked individuals. Males are consequently also grouped into four rank classes, from "high" to "very low."
37. Mahale: Nishida and Hosaka (1996); Kanyawara: Wrangham *et al.* (1992), Muller (2002).
38. Gombe: Goodall (1986: 418–423); Mahale: Nishida and Hosaka (1996); Taï N: Boesch and Boesch-Achermann (2000: 70–72); Budongo: Newton-Fisher (1999a); Ngogo: Mitani and Watts (2001), Mitani *et al.* (2002b).
39. Mitani (2009b: 217 and Fig. 2) discusses the heterogeneous nature of coalition formation among chimpanzee males, distinguishing those that reinforce dominance relationships ("conservative coalitions"), those that dramatically upset the dominance hierarchy ("revolutionary coalitions"), and those that help an individual improve their relative status within the hierarchy ("bridging coalitions").
40. Gombe: Goodall (1986: 435); Mahale: Nishida and Hosaka (1996: 123); Taï N: Boesch and Boesch-Achermann (2000: Table 6.4); Budongo: Newton-Fisher (2002: 134). The term "separating intervention" was coined by de Waal (1982) to describe observations at Burgers' Zoo (Arnhem, the Netherlands) of males disrupting the associations of other males.
41. Nishida (1983) coined the term "allegiance fickleness" to describe this case at Mahale. De Waal (1982) described in detail a similar situation at Burgers' Zoo. Duffy *et al.* (2007) found that the alpha male at Kanyawara interfered less in matings by those lower-ranking males that supported him frequently in aggressive interactions.
42. Goodall (1986: Table 12.4), based on seven Gombe males observed for 1,570 hours and seven females observed for 1,647 hours (Table 12.2).

43. Taï NS: Stumpf and Boesch (2010); Kanyawara: Muller *et al.* (2007); Budongo: Newton-Fisher (2006).
44. Attacks on stranger mothers and infants: cases at Gombe and Mahale prior to 1996 reviewed in Arcadi and Wrangham (1999: Table 5); for more recent incidents, see Kutsukake and Matsusaka (2002) at Mahale; Boesch *et al.* (2008) at Taï S; Watts *et al.* (2002) at Ngogo. Attacks on community mothers and infants: cases at Mahale, Kanyawara, and Budongo prior to 1996 reviewed in Arcadi and Wrangham (1999: Table 6); for more recent incidents, see Sakamaki *et al.* (2001) at Mahale, Newton-Fisher (1999b) at Budongo.
45. Gombe: Goodall (1986: Table 13.1) provides data on reassurance contact after aggression but without using the matched control methodology; Mahale: Kutsukake and Castles (2004) found similar reconciliation rates for males and females but used pooled data; Taï N: Wittig and Boesch (2005); Kanyawara: Muller (2002) found a higher rate of female–female than male–male reconciliation, but the number of female interactions was very small; Budongo: Arnold and Whiten (2001) observed no female–female reconciliation, but this was based on only three aggressive interactions.
46. Smuts (1986) proposed this explanation for sex differences in the frequency and severity of aggression based on her analysis of data from twenty-one primate species (Tables 32–1 and 32–2). See also Nishida (1989) for similar reasoning for chimpanzees.

7   SEXUAL BEHAVIOR: CONFLICTING STRATEGIES OF MALES AND FEMALES

1. Graham (1981), Wallis (1997).
2. Female chimpanzees normally nurse their infants for three to five years, during which time menstrual cycling ceases and males show little sexual interest in them. Normal cycling resumes soon after weaning or the death of an infant. Gombe: Goodall (1986: 483), Wallis (1997); Mahale: Nishida (1997), Nishida *et al.* (2003); Taï N: Boesch and Boesch-Achermann (2000); Kanyawara: Emery-Thompson (2007b); Budongo: Reynolds (2005).
3. Gombe: Goodall (1986); Mahale: Matsumoto-Oda *et al.* (1998), Matsumoto-Oda (1999a); Taï N: Anderson *et al.* (2002); Taï S: Wittiger and Boesch (2013); Kanyawara: Machanda *et al.* (2013); Budongo: Newton-Fisher (1999a); Ngogo: Mitani *et al.* (2002a).
4. Gombe: Goodall (1986: 445); Mahale: Hasegawa (1989); Bossou: Sugiyama and Koman (1979); Taï N: Boesch and Boesch-Achermann (2000:80); Kanyawara: Wrangham (2002).
5. Ngogo: Watts (2007: 226, 228).
6. Mahale: Matsumoto-Oda (1999b); estimates for Gombe, Mahale, Taï N, and Kanyawara in Wrangham (2002); Ngogo: Watts (2007).
7. Gombe: Goodall (1986: 450); Mahale: Takasaki (1985); Taï N: Boesch and Boesch-Achermann (2000: 80); Ngogo: Watts (2015).
8. Captive: Nadler *et al.* (1985); Taï S: Deschner *et al.* (2003); Kanyawara and Budongo: Emery Thompson (2005).
9. Gombe: Tutin (1979), Goodall (1986), Wallis (1997); Mahale: Hasegawa (1989), Nishida (1997), Matsumoto-Oda (1999a,b); Taï S: Deschner *et al.* (2004); Kanyawara: Emery Thompson (2005), Georgiev *et al.* (2014); Budongo: Emery Thompson (2005); Ngogo: Watts (2007).

10. Gombe: Tutin and McGinnis (1981) and Goodall (1986: 447–450) provide detailed descriptions of copulatory sequences; Mahale: Nishida (1997) provides detailed descriptions of male and female courtship gestures.

11. Gombe: In a five-year period, females failed to respond to male solicitation within one minute in only 4 percent (sixty-one) of 1,475 cases. Of these, two-thirds involved a sterile female or a sister, and sixteen times the female eventually mated (Goodall, 1986: 479). Mahale: In a two-year study, females refused courtship attempts in 17 percent of 229 cases, but most of these refusals were to two old adult males (Nishida, 1997). Taï N: In a two-month period, females avoided 8 percent of 123 male solicitations (Boesch and Boesch-Achermann, 2000: 80). In a subsequent twenty-eight-month study, females avoided copulating with males in 15 percent of 938 solicitations (calculated from Stumpf and Boesch, 2006: 755). Kanyawara: In an eleven-year period, females avoided only 3 percent of 1,894 copulation attempts (Gilby et al., 2010: 46).

12. For adult males at Gombe: mean of nine thrusts in seven seconds (Tutin and McGinnis, 1981: 248); Mahale: mean duration eight seconds (Hasegawa and Hiraiwa-Hasegawa, 1990), seven seconds (Nishida, 1997); Kanyawara: mean duration 6.6 seconds, one thrust per second (Wrangham, 2002).

13. Described by Tutin (1979), who also referred to possessive and consortship mating as "restrictive mating"; possessive mating appears to be absent at Taï, perhaps because there are few males remaining in the study groups, but aggressive interruptions of copulations are nonetheless observed there (Stumpf and Boesch, 2006: 755).

14. Gombe: Tutin (1979); Mahale: Hasegawa and Hiraiwa-Hasegawa (1983), Hasegawa (1989), Matsumoto-Oda (1999b); Taï N: Boesch and Boesch-Achermann (2000: 85); Kanyawara: Wrangham (2002); Budongo: Oliver (2002, in Reynolds, 2005: 118); Ngogo: Watts (1998).

15. Kanyawara: Muller and Wrangham (2004a), Georgiev et al. (2014).

16. Observed by Allen (1981) in a captive male.

17. Short (1979), Vahed and Parker (2012).

18. Tutin (1979); also referred to as "herding" (Nishida, 1997) and "mate guarding" (Watts, 1998).

19. Gombe: Tutin (1979: 32) reported a five-day maximum for possessive mating; Mahale: Hasegawa and Hiraiwa-Hasegawa (1983) observed possessive incidents involving the alpha male lasting up to seven days, while those involving lower ranking males only lasted up to one day; Nishida (1997: 387) reported a possible case of possessiveness lasting nine days; Ngogo: Watts (1998: 47) reported a maximum of four consecutive days of possessive mating but noted that most cases lasted only one day.

20. Gombe: Tutin (1979: 32); Mahale: Nishida (1979: 111), Hasegawa and Hiraiwa-Hasegawa (1983: Table 2, 1990: 126), Takasaki (1985: Table 2), Nishida (1997), Matsumoto (1999b); Kanyawara: Wrangham (2002: 211), Duffy et al. (2007); Budongo: Reynolds (2005: 118); Ngogo: Watts (1998).

21. Gombe: Tutin (1979), Tutin and McGinnis (1981); Mahale: Hasegawa and Hiraiwa-Hasegawa (1983), Takasaki (1985: Table 2); Kanyawara: Wrangham (2002), Muller et al. (2006); Budongo: Newton-Fisher et al. (2010); Ngogo: Sobolewski et al. (2013).

22. Gombe: Goodall (1986: 457) provides detailed descriptions of consortship formation behavior by various males.

23. Goodall (1986: 451) reports that consortships at Gombe can last up to three months.

24. Gombe: Tutin (1979), Goodall (1986: 464).

25. Gombe: Goodall (1986: 464) reported only four matings per day between a fully swollen female and male during consortship.
26. Gombe: For a twenty-two-year period, Wroblewski *et al.* (2009) identified seventeen conceptions by opportunistic, nine by possessive, and seven by consortship mating; Mahale: For a two-year period, Hasegawa and Hiraiwa-Hasegawa (1990) identified three by opportunistic, two by possessive, and one by consortship mating; Taï N: Boesch and Boesch-Achermann (2000: 77) reported only one certain conception from consortship in an eight-year period.
27. Gombe: Constable *et al.* (2001), Wroblewski *et al.* (2009); Mahale: Inoue *et al.* (2008); Bossou: Alpha male gets most copulations, Sugiyama and Koman (1979); Taï NMS: Boesch *et al.* (2006); Kanyawara: Alpha male gets most copulations, Georgiev *et al.* (2014); Budongo: Newton-Fisher *et al.* (2010); Ngogo: Alpha and other high-ranking males mate guard the most and get more copulations even when unable to prevent all other mating attempts (Watts, 1998).
28. Gombe: Goodall (1986: 445); Mahale: Hasegawa and Hiraiwa-Hasegawa (1983), Takahata (1990a), Nishida (1997), Matsumoto-Oda and Ihara (2011); Kanyawara and Budongo: Emery Thompson (2005).
29. Gombe: (Goodall, 1986: 477), Wrobloski *et al.* (2009), but note that Tutin (1979) reported that grooming and food sharing were better predictors of possessive mating; Mahale: Hasegawa and Hiraiwa-Hasegawa (1983), Takasaki (1985); Budongo: Reynolds (2005: 118); Ngogo: Watts (1998).
30. Gombe: Goodall (1986: 477, Tables 16.6, 16.7, E2, and E3), Wroblewski *et al.* (2009).
31. Mahale: Hasegawa and Hiraiwa-Hasegawa (1983: 77); Kanyawara: Wrangham (2002:212); Budongo: Reynolds (2005: 99); Ngogo: Watts (1998: 47).
32. Taï N: Boesch and Boesch-Achermann (2000: 80).
33. Taï NMS: Boesch *et al.* (2006).
34. Gombe: Tutin (1979), Goodall (1986: 483); Mahale: Nishida (1979: 118), who referred to estrous swellings as a "passport" for transferring females.
35. Kanyawara: Muller (2002), Muller *et al.* (2007, 2009); Budongo: Kaburu and Newton-Fisher (2015b). For general data on male aggression toward swollen females, see also Goodall (1986: Fig. 12.3) for Gombe; Matsumoto-Oda and Oda (1998: Table II) for Mahale. Note that with reduced male competition, for example if many females are swollen simultaneously or the male dominance hierarchy is more egalitarian, aggression may have a weaker influence on mating access. Mahale: Kaburu and Newton-Fisher (2015b).
36. Kanyawara: Muller *et al.* (2011).
37. Gombe: Feldblum *et al.* (2014).
38. Gombe: Tutin and McGinnis (1981: 261) reported two of 1,137 matings forced, one of which involved a mother and her son; Goodall (1986: 468).
39. Described by Goodall (1986: 482), Muller *et al.* (2009: 205).
40. This strategy was previously suggested for Gombe females by Tutin and McGinnis (1981: 263).
41. Gombe: In 1978, about 5 percent of male aggression overall was directed at females in sexual contexts, and about 10 percent of aggression received by adult females from adult males occurred in sexual contexts (Goodall, 1986: Figs.12.3a and 12.4c); Taï NS: 11.6 percent of sexual interactions involved aggression (Stumpf and Boesch, 2010).
42. Mahale: Nishida (1997); Taï NS: Stumpf and Boesch (2010).

43. Taï NS: Stumpf and Boesch (2005, 2006); Kanyawara: Pieta (2008) conducted a comparable analysis with similar results.
44. See Muller *et al.* (2011: 922) for discussion of how male aggression could confound the interpretation of female proceptivity and resistance to particular males.
45. Kanyawara: Emery Thompson *et al.* (2014).
46. "Sexual selection" refers to the selection of traits that increase mating success: "This depends on the advantage which certain individuals have over others of the same sex and species solely in respect of reproduction" (Darwin, 1871: Chapter VIII, paragraph 4). See Hrdy (1979; 1981: 153), Wrangham (1993), and van Schaik *et al.* (1999) for discussions of paternity confusion as a defense against infanticide.
47. Wallis (1992).
48. Boesch and Boesch-Achermann (2000: 58).
49. Ngogo: Watts (2007: 230); see also Hashimoto and Furuichi (2006) for high copulation rates in the Kalinzu (Uganda) community, which has a large number of males.
50. Goodall (1986: 475) suggested that during consortships away from group mates, males risk losing dominance status and make themselves vulnerable to intercommunity aggression; Georgiev *et al.* (2014) found that Kanyawara males spent less time feeding, associated with giving and receiving more aggressive acts, in the presence of swollen females.
51. See also Wrangham (2002) for a discussion of factors producing variation in female copulation rates over sexual cycles.

8   COALITIONARY LETHAL AGGRESSION BETWEEN AND WITHIN COMMUNITIES

1. Mech *et al.* (1998).
2. Wrangham (1999).
3. There are numerous published accounts of the behavioral details of fatal attacks in chimpanzees, many of which are endnoted in this chapter. The case study descriptions from Goodall *et al.* (1979) and Goodall (1986: Chapter 17) provide a wide range of representative examples involving different classes of victims in various contexts.
4. Wilson *et al.* (2014a).
5. Gombe: Williams *et al.* (2008).
6. Wrangham *et al.* (2006: Tables 4, 5, and 6) calculated median rates of 271, 164, and 595 deaths per 100,000 individuals per year for chimpanzee, hunter-gather, and subsistence farming groups, respectively.
7. Goodall (1986: 529).
8. For example, see Wrangham (1999) and Wilson and Wrangham (2003) versus Marks (2002) and Ferguson (2011).
9. Wilson *et al.* (2014a). In this study, the authors restricted their statistical analyses to deaths that were directly observed or inferred, and discounted those that were only suspected. They used the following criteria: "We rated a case as observed if observers directly witnessed the attack. We rated a case as inferred if the attack was not directly witnessed, but compelling evidence indicated that the victim was killed by chimpanzees (such as a body found with multiple bite wounds, and/or skeletal trauma consistent with a chimpanzee attack). We rated other cases as suspected; for example,

disappearances of chimpanzees that appeared healthy before their disappearance (with the exception of adolescent females, who generally disperse from their natal community), or individuals known to have died from wounds that may have been inflicted by chimpanzees."

10. For example, at Gombe, Mjungu (2010) reported that of 225 intercommunity encounters, 210 were vocal, ten visual, and five physical; similarly at Mahale, Nishida (1979: 85) reported a priority of access relationship between K-group and M-group in which K-group always avoided M-group once its presence was detected.

11. Gombe: Bygott (1979), Goodall *et al.* (1979); Mahale: silent travel and scouting reported in Nishida (1979: Table 3); Tai N: Boesch and Boesch-Achermann (2000: 136); Kanyawara: Wrangham (1999: 7); Budongo: Reynolds (2005: 107), Bates and Byrne (2009: 253); Ngogo: Watts and Mitani (2001).

12. Gombe: Goodall (1986: Table 17.1); Ngogo: Mitani and Watts (2005: Fig. 1).

13. Gombe: Gilby *et al.* (2013a); Taï N: Boesch and Boesch-Achermann (2000: Table 7.9); Kanyawara: Chapman and Wrangham (1993) reported that males are seen more often in boundary areas than females, and Wilson *et al.* (2007) found that parties observed in border areas included more males; Budongo: Bates and Byrne (2009) found that males were more likely to travel in border areas; Ngogo: Mitani and Watts (2005).

14. Gombe: Goodall *et al.* (1979), Goodall (1986), Wilson *et al.* (2004); Mahale: Nishida *et al.* (1985); Taï N: Boesch and Boesch-Achermann (2000: 136), Boesch *et al.* (2008); Wrangham (1999: 8); Budongo: Reynolds (2005: 107) reported suspected cases of lethal aggression beyond community borders; Ngogo: Mitani *et al.* (2010).

15. Manson and Wrangham (1991), Wrangham (1999).

16. Wilson *et al.* (2001).

17. Manson and Wrangham (1991), Wrangham (1999).

18. Taï N: Boesch (1996).

19. Taï NMSE: Boesch *et al.* (2008).

20. Mahale: Nishida (1968: 204), Kawanaka and Nishida (1974).

21. Manson and Wrangham (1991), Wrangham (1999).

22. The events at Gombe briefly summarized in this section are described in detail by Goodall (1986: 503–514).

23. The events at Mahale briefly summarized in this section are described in detail by Nishida *et al.* (1985).

24. Mahale: Hiraiwa-Hasegawa *et al.* (1984: Table 1).

25. Mahale: Nishida (1979).

26. Mahale: Nishida *et al.* (1979).

27. The events at Ngogo outlined in this section are reported in Mitani *et al.* (2010).

28. Gombe: The daughter of the Kahama female killed by Kasekela males subsequently transferred to the Kasekela community (Goodall, 1986: 514); Mahale: Four K-group females transferred to M-group after the K-group males disappeared (Nishida, 1979: 298); Ngogo: Three parous females immigrated into the community after the spate of killings and territorial expansion to the northeast (Wood *et al.*, 2017: 44).

29. Gombe: Williams *et al.* (2002a).

30. Gombe: Goodall (1986: Table 17.2), Williams *et al.* (2004), Wilson *et al.* (2004); Mahale: Nishida and Hiraiwa-Hasegawa (1985); Taï N: Boesch and Boesch-Achermann (2000: 156); Taï NMSE: Boesch *et al.* (2008); Kanyawara: Arcadi and Wrangham (1999), Muller (2002: 118); Budongo: Newton-Fisher (1999c); Ngogo: Watts and Mitani (2001), Mitani *et al.* (2010).

31. Reviewed in Lee (1987), Atsalis and Videan (2009).
32. Williams *et al.* (2004) reported lower interbirth intervals at Gombe with increased territory size; analyzing thirty-three years of Gombe data, from 1967 to 2000, Pusey *et al.* (2005) found that female weights were highest in the early years of heavy provisioning. Between 1973 and 2000, female weights were also heavier when range size was greater and population density was lower.
33. Wilson *et al.* (2014a: Extended Data Table 3): five observed, five suspected, and five inferred.
34. Gombe: Goodall (1986: 89, 111).
35. Gombe: Goodall (1992).
36. Nishida (1983).
37. Nishida *et al.* (1995).
38. Gombe: Goodall (1992); Mahale: Nishida (1983), Kaburu *et al.* (2013); de Waal (1982/1998) reported a similar case in captivity.
39. See Wilson *et al.* (2014a: Extended Data Table 3) references for younger and older males. No information is provided for the deaths of the other prime-aged males.
40. Suzuki (1971).
41. Wilson *et al.* (2014a: Extended Data Tables) provide the most recent tallies. For additional reviews, see Hamai *et al.* (1992), Arcadi and Wrangham (1999), Murray *et al.* (2007b).
42. For example, Arcadi and Wrangham (1999) at Kanyawara.
43. Takahata (1985), Nishida and Kawanaka (1985).
44. Wilson *et al.* (2014a: Extended Data Table 2).
45. Watts *et al.* (2002).
46. As discussed in Chapter 7, it is widely assumed that female promiscuity evolved as a counter-strategy against intracommunity male infanticide (Hrdy, 1977). In at least one infanticide case at Mahale, the killers were suspected to have mated with the mother during the cycle in which the infant was conceived: Nishida and Kawanaka (1985).
47. This sexual selection–based hypothesis closely resembles the widely accepted explanation for infanticide in langur monkeys and gorillas, species in which a single male can maintain exclusive mating access to several females until replaced by an outsider male, who may then kill young infants and mate with their mothers (Hrdy, 1974, 1979; Watts, 1989).
48. Mahale: Takahata (1985), Nishida and Kawanaka (1985), Sakamaki *et al.* (2001); Kanyawara: Arcadi and Wrangham (1999).
49. Mahale: Hamai *et al.* (1992).
50. Gombe: Goodall (1977; three observed, one inferred), Pusey *et al.* (2008b; one inferred); Taï N: Boesch and Boesch-Achermann (2000: 33; one suspected, based on females found eating an infant); Budongo: Townsend *et al.* (2007; two inferred); unpublished data from the Budongo Chimpanzee Project reported in Wilson *et al.* (2014a; one observed, one inferred).
51. Gombe: Pusey *et al.* (2008b).
52. Goodall (1977), Hiraiwa-Hasegawa (1992).
53. Pusey (1983: 376).
54. Reviewed in Vasey (2006: 277).
55. Power (1991).
56. Wilson *et al.* (2014a: 415).
57. For example, Struhsaker (1999).
58. An unusual influx of either stranger or peripheral adult females into the Sonso community at Budongo was linked to infanticidal attacks by resident

adult females and was possibly a result of habitat disturbance: Emery
Thompson *et al.* (2006), Townsend *et al.* (2007), and Langergraber *et al.*
(2014). See also debate between Ferguson (2014) and Wilson *et al.* (2014b)
over the role of human impacts on chimpanzee lethal aggression.

59. I am indebted to Bill Wallauer, who described to me the flight of unhabituated chimpanzees from patrolling Ngogo males followed by himself and
other human observers, for alerting me to the possible impact of researchers on territorial encounters.

9   HUNTING, EATING, AND SHARING MEAT

1. Hunting is often not observed in the early stages of field projects, before
study animals are well habituated and can be continuously tracked (e.g., Taï
N: Boesch and Boesch, 1989: 549; Budongo: Reynolds, 2005: 73). It is probably less common at Budongo because the preferred prey of chimpanzees
elsewhere, red colobus monkeys, are absent there (Plumptre and Reynolds,
1994). It appears to be extremely infrequent at Bossou because prey species
in general are rare there (Hirata *et al.*, 2001).

2. Gombe: about four a month (calculated from Stanford *et al.*, 1994a: 529 hunts
in eleven years); Mahale: about five a month (calculated from Hosaka *et al.*,
2001: 117 hunts in two years); Taï N: 4.5 or ten a month (Boesch and Boesch,
1989: 555); Ngogo: four a month (Mitani and Watts, 1999).

3. Reviewed in Gilby *et al.* (2017).

4. Gombe: Busse (1977: 908), unpublished data cited in Stanford (1996: 99).
Several estimates of average per capita meat consumption, based on capture
and consumption rates, have also been attempted. In most cases, substantial
individual variation is documented. Gombe: 25 and 8 kg/yr for adult males
and adult females, respectively (Wrangham and Bergmann Riss, 1990: 167);
Mahale: 16.8 and 16.4 kg/yr, with alpha males obtaining five times as much
as other males (Hosaka, 2002, in Hosaka, 2015a: 282); Taï N: 67.9 and 9.1 kg/
yr (Boesch and Boesch-Achermann, 2000: Table 8.4); Ngogo: 15–20 (range 2–
40) and 3.7 (up to 10) kg/yr (Watts and Mitani, 2002a: 17).

5. Reviewed in Tennie *et al.* (2009a).

6. See Teleki (1973: 126–135) and Goodall (1986: 270–285) for detailed descriptions of hunting techniques for different types of prey at Gombe.

7. Taï N: Chimpanzees are deemed to "search" for prey if they change travel
direction without seeing or hearing prey, "become totally silent, remain very
close together, move one behind the other, and stop regularly to look up into
the trees, alert to the sound of monkeys." Between 1984 and 1986, thirty-nine
of seventy-eight (50 percent) hunts were preceded by searching (Boesch and
Boesch, 1989: 554). Ngogo: Between 1994 and 1998, twenty of forty-nine (41
percent) hunts followed searches (Mitani and Watts, 1999: 445). From 1998 to
2012, 171 of 418 (40.9 percent) encounters with red colobus monkeys followed hunting patrols at Ngogo (Watts and Amsler, 2013: 932).

8. Wrangham (1975: 4.36) and Goodall (1986: 291) describe cases of prolonged
killing episodes at Gombe.

9. Gombe: Goodall (1968a: 191); Goodall (1986: 297) also cites an example of a
prey braincase being wiped clean with leaves; Mahale: Uehara *et al.* (1992:
150); Taï N: Boesch and Boesch (1989: 562).

10. Struhsaker (2010: Appendix 3.2).

11. Gombe: Busse (1977), Goodall (1986: 274), Stanford (1998: 152 and Fig. 7.5).

12. Ngogo: Watts and Mitani (2015: 735).

13. Gombe: Aimed throwing to intimidate or scare off other animals is common, however, and Goodall (1986: 554) reports incidents of chimpanzees throwing rocks at adult pigs and baboons defending their young (see also Plooij, 1978a); Mahale: Huffman and Kalunde (1993) report a case of a twelve-year-old female using a stick to "rouse" a squirrel from its nest, which she subsequently grabbed and killed; Taï N: Boesch and Boesch (1989: 565) report a case of a male chimpanzee throwing a branch at adult colobus monkeys that were threatening him.
14. Tool-assisted hunting at Fongoli is described in Pruetz and Bertolani (2007).
15. As of 2015, researchers at Fongoli recorded 308 tool-assisted hunts, of which the hunter's identify was known in 294 cases. Over a ten-year period, 21 percent of successful hunts involved tool use. Of hunts in which the sex of the hunter was known, 56.8 percent were by females despite their being underrepresented compared with males in parties on hunting days. Although individuals of all ages and both sexes hunted with tools, adult males and females were much more successful, capturing their prey about 20 percent of the time (Pruetz et al., 2015).
16. See Pruetz (2006) for general feeding ecology of Fongoli chimpanzees.
17. Gombe: Wrangham (1974).
18. Gombe: Goodall (1986: 292); see Gilby et al. (2017) for recent observations of "piracy" by both Kasekela and Mitumba chimpanzees.
19. Taï N: Boesch and Boesch-Achermann (2000: 170).
20. Gombe: fewer than twenty scavenging cases total in thirty-four years (Goodall, 1986: 295; Stanford, 1996: 97), three additional cases reported in Gilby et al. (2017); Mahale: seven cases in twenty-five years (Uehara, 1997: 195); Taï N: seven cases in twenty years (Boesch and Boesch-Achermann, 2000: 170); Kanyawara: one case in twenty years (Gilby et al., 2017); Budongo: one case in fifteen years (Reynolds, 2005: 74); Brand et al. (2014) also reported two cases from Budongo of chimpanzees preying on blue duikers injured in snap-traps set by poachers; Ngogo: four cases in thirteen years (Watts, 2008).
21. Mahale: Hosaka et al. (2001: Fig. 23), Hosaka (2015a: Table 20.1).
22. Ngogo: Watts and Amsler (2013), Watts and Mitani (2015).
23. Gombe: Stanford (1998: 112).
24. Gombe: Gilby et al. (2006: Figure 1, Table 4) for 1974–2001; see also Stanford et al. (1994a) for a similar result using early Gombe data subsumed by Gilby et al. (2006); Mahale: Hosaka et al. (2001: Table 10) for party size, and Hosaka (1998, cited in Hosaka, 2015a: 283) for number of males; Taï N: Boesch and Boesch-Achermann, 2000: Table 8.10) for party size; Kanyawara: Gilby and Wrangham (2007: 1776), Gilby et al. (2008: Fig. 1), Gilby et al. (2015: Fig. 1); Ngogo: Mitani and Watts (2001).
25. Watts and Mitani (2002b: 252).
26. This is especially evident at Ngogo, where the average size of hunting parties and the number of males in them are both unusually high, and the rates of both hunting episodes/encounter and success/hunting episode greatly exceed those of other communities (Mitani and Watts, 1999: Tables 4 and 7). At Gombe, Stanford et al. (1994b: 225) concluded that increased hunting rates in the 1980s were a consequence of an increased number of mature males in the community available to form hunting parties.
27. Kanyawara: Gilby and Wrangham (2007); Ngogo: Mitani and Watts (2001). At Gombe, Gilby et al. (2006) found no effect of diet quality (percentage of time spent feeding on leaves or pith versus fruits) on hunting probability, but this study was not based on phenological data.

28. No difference in predation rates between dry and rainy seasons: Wrangham (1975: 4.8) and Wrangham and Bergmann Riss (1990) at Gombe. More hunting in the dry season: Takahata *et al.* (1984) at Mahale. More hunting in the late dry season and early rainy season: Stanford *et al.* (1994b) at Gombe, Hosaka *et al.* (2001) at Mahale. More hunting in the rainy season: Boesch and Boesch-Achermann (2000: 161) at Taï N.

29. For studies of fruiting phenology at the long-term field sites, see the following. Mahale: Itoh and Muramatsu (2015); Taï N: Anderson *et al.* (2005), Polansky and Boesch (2013); Kanyawara: Chapman *et al.* (1999, 2005, 2011); Budongo: Babweteera *et al.* (2011); Ngogo: Chapman *et al.* (1999), Watts *et al.* (2012b). Note that at Gombe, Pintea *et al.* (2011) documented habitat changes with satellite imagery and inferred changes in fruit abundance from changes in time spent feeding by chimpanzees on different fruit species.

30. Gombe: Stanford *et al.* (1994a); Mahale: Takahata *et al.* (1984); Budongo: Suzuki (1971, in Nishida *et al.*, 1979: 15).

31. Ngogo: Mitani and Watts (2001), Watts and Mitani (2002a, 2015). Interestingly, positive correlations between food abundance and hunting are also consistent with descriptions from the earliest detailed study of hunting at Gombe, which suggested that hunting was not driven by hunger, since it often occurred after sessions of intensive banana feeding at the provisioning station (Teleki, 1973: 109).

32. Ngogo: Mitani and Watts (1999), Amsler (2010).

33. Boesch (1994: 655).

34. Stanford *et al.* (1994a).

35. Gombe: Gilby *et al.* (2006: Fig. 3), although in a more recent analysis Gilby *et al.* (2015: 11) found that the presence of sexually receptive females had no effect, either positive or negative; Kanyawara: Gilby *et al.* (2015: 11); Ngogo: Mitani and Watts (2001: 919). Note that in the one study reporting that the presence of swollen females increased the probability of hunting (Stanford *et al.*, 1994a, at Gombe), partially swollen females that males rarely mate with were included in the analysis, and "hunting" included "interest" in hunting rather than being restricted to the more stringent requirement of active pursuit (see Gilby *et al.*, 2006: 176, who also critique the statistical methods employed by Stanford *et al.*).

36. Gombe: Wrangham (1975: 4.11).

37. Gombe: Stanford (1994a: 12), Gilby *et al.* (2006); Ngogo: Watts and Mitani (2002a).

38. Stanford (1998: Figs. 6.1–6.3).

39. Ngogo: Mitani and Watts (1999: 451) report that patrolling chimpanzees do not hunt when they encounter red colobus groups from which they have recently "harvested" young.

40. Several observers have reported the subjective impression that chimpanzees are not cooperating. Gombe: Busse (1978), Stanford (1994b: 226); Mahale: Nishida *et al.* (1979: 19).

41. See Packer and Ruttan (1988) for mathematical modeling of payoffs in cooperative hunting.

42. Gombe: Stanford (1996: 101) (no data provided), Gilby *et al.* (2006); Taï N: Boesch (1994: Fig. 2) found that the potential individual payoff in kilojoules was greater with three to five than with one or two hunters, but then declined substantially in larger hunting parties; Ngogo: Watts and Mitani (2002a).

43. Packer and Ruttan (1988), Watts and Mitani (2002a).

44. Gombe: Tennie *et al.* (2009a); Kanyawara: Gilby *et al.* (2008).

45. Teleki (1973: 132) described in general terms the "considerable coordina-tion" used by Gombe hunting parties stalking a prey group. Boesch and Boesch (1989), Boesch (1994a), and Boesch and Boesch-Achermann (2000) proposed an elaborate scheme of hunting cooperation among Taï N chim-panzees based on qualitative visual assessment, involving specific indivi-dual hunting roles and levels of complexity in the coordination of these roles. Researchers elsewhere have reported the difficulty or impossibility of collecting systematic data to examine this proposal – Gombe: Busse (1978), Stanford (1996: 100); Mahale: Hosaka *et al.* (2001: 117); Ngogo: Watts and Mitani (2002a: 5).
46. Gombe: Gilby *et al.* (2015: 9), based on rates of group hunt initiations, identified three males over the course of thirty-seven years in the Kasekela community; Mahale: Hosaka (2015a: 285), citing Hosaka *et al.* (2001), sug-gested that skillful hunters may catalyze group hunting, but no data on hunting initiations are provided; Taï N: Boesch and Boesch (1989: 556) identified one eager male hunter, and Boesch and Boesch-Achermann (2000: 180) identified a second male that frequently initiated hunts, but no data provided; Kanyawara: Gilby *et al.* (2015: 9), based on rates of group hunt initiations over eighteen years, identified two males; Ngogo: Watts and Mitani (2002a: 22) identified two males that seemed especially motivated to hunt, but no data provided.
47. Gombe and Kanyawara: Gilby *et al.* (2015: 9); Taï N: Boesch and Boesch (1989: 556).
48. For detailed descriptions from Gombe of the variety of behavioral events that can follow prey capture, see Teleki (1973: 135–151) and Goodall (1986: 289, 299–312, 372–374); Taï N: Boesch and Boesch (1989: 560–565).
49. Gombe: Busse (1978), Goodall (1986: 306 and Table 11.12).
50. Gombe: Teleki (1973: 148); Budongo: Reynolds (2005: 75).
51. Gombe: Wrangham (1975: 4.37), Busse (1978: 769); Mahale: Nishida *et al.* (1979: 19).
52. But note that mothers do share with offspring. Gombe: Teleki (1973: Fig. 12); Mahale: Zamma (2005); Budongo: Reynolds (2005: 75).
53. The term "sharing-under-pressure" was coined by Gilby (2006), but the hypothesis was previously articulated and explored by Wrangham (1975: 4.53–4.59).
54. Wrangham (1975: 4.57).
55. Gombe: Gilby (2006).
56. Meat sharing between males and females could in principle function to build social bonds, but more often appears to result from especially persis-tent begging by females – Gombe: Teleki (1973: 162–163), Gilby (2006); Mahale: Nishida *et al.* (1992) report several instances of sharing with older females but provide no information about begging; Taï S: Gomes and Boesch (2009) report many instances of male-to-female meat transfer but do not report the nature of female begging behavior; Budongo (for plant-food sharing): Slocombe and Newton-Fisher (2005).
57. Mahale: Nishida *et al.* (1992).
58. Ngogo: Watts and Mitani (2002b).
59. Taï S: Gomes and Boesch (2011); Ngogo: Mitani and Watts (2001), Watts and Mitani (2002b).
60. Gombe: Gilby (2006).
61. Budongo: Wittig *et al.* (2014a).
62. Gombe: Teleki (1973: 162–163).

63. Stanford (1996: 101). Note also that although Gilby (2006: 959) found that one male possessor was more likely to share with female grooming partners, these females were also the most persistent beggars.
64. Gombe: Gilby (2006), Gilby *et al.* (2010); Kanyawara: Gilby *et al.* (2010); Ngogo: Mitani and Watts (2001), Watts and Mitani (2002b).
65. Taï S: Gomes and Boesch (2009).
66. Stanford (1995: Table I).
67. Stanford (1998: 187, Appendix 22).
68. Lwanga *et al.* (2011), based on 372 surveys on the same line transect over the course of 32.9 years. See also Mitani *et al.* (2000b) for a detailed examination of census methods at Ngogo.
69. Watts and Mitani (2015: Fig. 1); see also Watts and Amsler (2013: Fig. 2).
70. Gombe: Fourrier *et al.* (2008); Ngogo: Teelen (2008).
71. Ngogo: Watts and Amsler (2013: 935).
72. Struhsaker (1999).
73. Watts and Amsler (2013), Watts and Mitani (2015).
74. For example, see contributions in Muller *et al.* (2017).

10   COMMUNICATION: THE FORM AND CONTENT OF SIGNALS

1. "Signals" are distinguished from "cues," information detected by an animal's sensory organs that allows the animal to evaluate conditions in the external environment. Information about the state of another animal is considered a cue if it is generated inadvertently, for example a physical sign of aging or poor health (Bradbury and Vehrencamp, 2011: 4).
2. Smith (1965, 1969, 1977: 19). See Fischer and Price (2016) for a review of recent theoretical developments in the analysis of primate signals.
3. Gombe: Goodall (1986: 137); Mahale: Nishida (1997), Matsumoto-Oda *et al.* (2007).
4. Gombe: Goodall (1967/2005: 304).
5. Goodall (1986: 122).
6. Discriminating between "levels of intentionality" provides a conceptual framework for characterizing the mental states of animals during communicative interactions. In first-order intentionality, an animal produces a signal with the intention of causing another animal to behave in a certain way, irrespective of the knowledge state of the receiver. For example, a monkey might produce an alarm call only because it sees a predator, thereby causing a group mate to flee. In second-order intentional communication, evidence for which is inconclusive in primates, an animal would have an understanding of the knowledge state of its own and others' minds. Thus in a theoretical case of alarm calling, the caller believes that the listener is ignorant of the predator's presence and calls in order to provide information to change the receiver's mental state, thereby causing him or her to flee. Reviewed in Cheney and Seyfarth (2007: 147), following Dennett (1987).
7. Huchard *et al.* (2009) for wild chacma baboons (*Papio ursinus*).
8. Fitzpatrick *et al.* (2015) for wild savannah baboons (*Papio cynocephalus*).
9. Nunn (1999). Deschner *et al.* (2004) provide data from Taï S supporting the graded-signal hypothesis for chimpanzees.
10. The "play face," or "relaxed open-mouth face," frequently occurs with laughing but does not appear to be a consequence of the acoustic demands of the vocalization. Rather, the two signals occur together more often when

play involves physical contact, suggesting that their combined production reflects elevated arousal (Davila-Ross *et al.*, 2015).

11. For example, "full closed grin" ("silent bared teeth display") with smiling in humans, play face with laughing in humans, and "compressed-lips face" with a human "angry face" (van Hooff, 1967; Waller and Dunbar, 2005; Parr and Waller, 2006; Parr *et al.*, 2007).

12. Burgers' Zoo (Arnhem, the Netherlands): de Waal (1982/1998: 37).

13. Tomasello and Call (2007: 8); Hobaiter and Byrne (2011a: 749).

14. Budongo: Hobaiter and Byrne (2011a), Roberts *et al.* (2014).

15. Taï S and Kanyawara: Fröhlich *et al.* (2016).

16. Gombe: Goodall (1986: 581).

17. Burgers' Zoo (Arnhem, the Netherlands): de Waal (1982/1998: 34).

18. Gombe: Plooij (1978b); Taï S and Kanyawara: Fröhlich *et al.* (2017); captive studies: Tomasello *et al.* (1994), Call and Tomasello (2007).

19. Hobaiter and Byrne (2011a) reported no gestures unique to a single individual at Budongo. They also compare (Table 1) the sixty-six gestures they documented in their study with sixty-one reported from Gombe and sixty-nine from Mahale; Nishida *et al.* (2010: 21–22) listed sixty-nine gestures in their species-wide ethogram. Roberts *et al.* (2014) identified 120 gesture types at Budongo but note (p. 333) that the increase compared to Hobaiter and Byrne (2011a) reflects a methodological difference in classifying graded signals with multiple components.

20. Budongo: Hobaiter and Byrne (2011b), Roberts *et al.* (2013).

21. Marler and Tenaza (1977: Table 16).

22. Mahale: Mitani and Nishida (1993: Fig. 1); Kanyawara: Clark (Arcadi) (1993: Table IV, Appendix A), Arcadi (2000), Wilson *et al.* (2007), Fedurek *et al.* (2014: Fig. 3).

23. Gombe: Wrangham (1975: 5.50); Mahale: Mitani and Nishida (1993); Kanyawara: Fedurek *et al.* (2014); Budongo: Notman and Rendall (2005).

24. Gombe: Marler and Hobbet (1975); Mahale: Mitani and Brandt (1994), Mitani *et al.* (1996); Taï NMS: Crockford *et al.* (2004); Budongo: Notman and Rendall (2005).

25. Primate Research Institute, Kyoto University: Kojima *et al.* (2003).

26. Kanyanchu (Kibale National Park): Levréro and Mathevon (2013).

27. Budongo: Townsend *et al.* (2011).

28. Gombe: Plooij (1980); Primate Research Institute, Kyoto: Kojima (2001).

29. Gombe: Marler and Hobbett (1975).

30. Mahale: Mitani and Gros-Louis (1995). Note, however, that this study was conducted before more recent analyses that have suggested the existence of functionally significant "scream" variants, the existence of which could potentially have confounded the earlier analysis of sex differences (Table 10.3 in this volume).

31. Mahale: Hayaki (1990), Sakamaki (2011), Sakamaki and Hayaki (2015); Budongo: Laporte and Zuberbühler (2010).

32. Mahale: Mitani and Nishida (1993); Kanyawara: Clark (Arcadi) (1993), Arcadi (1996, 2000), Wilson *et al.* (2007), Fedurek *et al.* (2014).

33. Kanyawara: Clark (Arcadi) and Wrangham (1994).

34. Males at Kanyawara sometimes abruptly moved a meter or two during the climax phase, giving the subjective impression of elevated arousal (personal observation). Similarly, when produced along with "pant hoots," "buttress drumming" often occurs following the build-up phase or during the climax (see text). It is interesting to note that in their study of calling behavior in captive chimpanzees, Kojima *et al.* (2003: 226; Primate Research Institute,

Kyoto University) reported that one individual typically did not produce climax elements but instead threw stones at this point in the call.

35. Kanyawara: Riede *et al.* (2004, 2007).
36. Kanyawara: Fedurek *et al.* (2016).
37. Taï NMSE: Boesch *et al.* (2008).
38. Gombe versus Mahale: Mitani *et al.* (1992); Ngogo versus Mahale: Mitani *et al.* (1999); Taï NMS: Crockford *et al.* (2004).
39. Kanyawara versus Gombe and Mahale: Arcadi (1996); Ngogo versus Mahale: Mitani *et al.* (1999).
40. Kanyawara: Wilson *et al.* (2001).
41. Taï NMS: Herbinger *et al.* (2009).
42. Kanyawara: Wilson *et al.* (2007).
43. A study at Taï NM reported that pant hoots, screams, and drumming were produced in combination more frequently than alone (Crockford and Boesch, 2005). In an unusual methodological departure, however, these authors treated the climax phase of pant hoots as a separate scream vocalization, although pant hoot climax elements are acoustically different than screams uttered in aggressive contexts (Riede *et al.*, 2007). This methodological choice artificially inflated the proportion of combined calls.
44. See Arcadi (2005) for a brief summary of the distinctions between words and animal calls.
45. Marler *et al.* (1992), Macedonia and Evans (1993).
46. As originally described, a vocalization can be deemed "functionally referential" if it meets two criteria: (1) It is only produced in the presence of a particular stimulus ("production specificity"), and (2) playbacks elicit the same behavioral response that listeners exhibit when hearing the call under natural conditions ("response specificity"). See Wheeler and Fischer (2012, 2015) and Fischer and Price (2016) versus Townsend and Manser (2013) and Scarantino and Clay (2015) for recent debate over the efficacy of the concept.
47. Kanyawara: Arcadi (unpublished data).
48. Gombe: Marler (1969), Marler and Tenaza (1977: 1024).
49. Gombe: Goodall (1968b: Table 12–2), Wrangham (1975: 3.32).
50. Edinburgh Zoo, Scotland: Slocombe and Zuberbühler (2005b).
51. Taï S: Kalan *et al.* (2015).
52. For example, alarm calls in vervet monkeys (*Chlorocebus pygerythrus*) (Seyfarth *et al.*, 1980). But see Price *et al.* (2015) for evidence that specificity in even this classic case is less than initially thought.
53. Budongo: Slocombe and Zuberbühler (2006).
54. Arcadi (2005).
55. Gombe: Wrangham (1975: 3.31); Kanyawara: Arcadi (personal observation).
56. Gombe: Goodall (1986: 125).
57. Kanyawara: Fedurek and Slocombe (2013); Budongo: Slocombe *et al.* (2010), Schel *et al.* (2013).
58. Mahale: Mitani and Brandt (1994), Mitani and Gros-Louis (1998); Kanyawara: Fedurek *et al.* (2013a, b); Budongo: Fedurek *et al.* (2013b).
59. Gombe: Arcadi and Wallauer (2013); Tai N: Arcadi *et al.* (1998: Fig. 4); Kanyawara: Arcadi *et al.* (2004: Fig. 2); Budongo: Babiszewska *et al.* (2015: Table 4). The longest bout in Arcadi and Wallauer (2013) was twenty-six beats in 5.5 seconds.
60. Boesch (1991b) speculated that one male at Taï N temporarily produced different types of drum bouts to relay different messages about his travel intentions. However, it was not possible in this study to control for the possibility that listeners were able to determine the drummer's travel

direction through simple triangulation (Arcadi, 2000: 219). In addition, listeners were never systematically monitored, and this male reportedly stopped exhibiting the behavior before the claim could be confirmed.

61. Galloping is an asymmetrical gait pattern in which the footfalls are not evenly spaced in time, in contrast to symmetrical gaits such as walking and running. In galloping, the hind limbs make ground contact in quick succession, followed by a longer interval during which they are in the air before their next closely timed ground contact. In complete stride cycles, hind limb and forelimb couplets alternate in this fashion (Hildebrand, 1977). Goodall (1968a) was the first to describe the use of the feet to produce the double beat pattern. Arcadi and Wallauer (2011, 2013) used audiovisual records from Gombe to illuminate the connection between gallop-gait limb sequences and drum beat timing.

62. Budongo: Fedurek *et al.* (2015b), based on about seventeen hours of data from 192 grooming bouts.

63. Ngogo: Pika and Mitani (2006), based on the groomer attending to the scratched spot in 119 of 186 cases (64 percent).

64. For example, see Cheney and Seyfarth (2007: Chapter 11) for a discussion of how the evolution of language may have involved the recruitment of cognitive mechanisms already in place to navigate complex social interactions.

65. Mitani *et al.* (1992, 1999).

66. See Marshall *et al.* (1999) (Lion Country Safari, USA) for adoption of an introduced pant hoot variant that included a "raspberry" sound produced by forcing air though the lips, Hopkins *et al.* (2007) (Yerkes National Primate Research Center) for the attention-getting use of "raspberry" and extended grunt sounds, and Watson *et al.* (2015a) (Edinburgh Zoo, UK) for acoustic convergence in the pitch of food grunts after the merger of two captive groups. Note, however, that wild chimpanzees also produce sounds by forcing air through the lips, and also utter drawn out grunts ("splutter" and "grunt, extended" in Nishida *et al.*, 2010). In addition, as previously discussed, it is difficult to dismiss the role of arousal in food grunt variation (Fischer *et al.*, 2015, Watson *et al.*, 2015b), especially when only a tiny fraction of calls produced in call bouts are examined.

67. Kanyawara: Clark (Arcadi) (1993), Clark (Arcadi), and Wrangham (1994).

68. Kanyawara: Wilson *et al.* (2001, 2007).

69. Budongo: Slocombe and Zuberbühler (2007), Fedurek *et al.* (2015a).

70. Budongo: Townsend *et al.* (2008), Fallon *et al.* (2016).

71. Budongo: Laporte and Zuberbühler (2010).

72. Budongo: Schel *et al.* (2013).

73. Studies of mental state attribution in captive chimpanzees have generated conflicting results. For a recent review, see Andrews (2017).

74. Budongo: Gruber and Zuberbühler (2013) for travel hoos.

75. Schel *et al.* (2014) for alarm huus and barks.

11   COMMUNITY DIFFERENCES IN GROOMING POSTURES AND TOOL USE: INNOVATION, SOCIAL LEARNING, AND THE QUESTION OF "CULTURE"

1. Whiten *et al.* (1999: 683, 2001). See Goodall (1973: Tables II and III) for a groundbreaking attempt to identify community-level variation in

chimpanzee tool use and social behavior, drawing on published reports from early field studies across Africa.

2. Subsequent analyses have sought to determine whether genetic factors could be implicated in generating patterns of variation, with mixed results. Lycett *et al.* (2007, 2009, 2010) found that cladistic analyses of behavioral variants did not match the phylogenetic tree for the species and concluded that social learning best explained intercommunity behavioral variation. Langergraber *et al.* (2011) and Langergraber and Vigilant (2011) found that behavioral and genetic similarity between communities were correlated and concluded that genetic factors could not be excluded. Kamilar and Marshack (2012) found that behavioral similarity increased with geographic proximity, supporting the idea that female dispersion is implicated in behavioral diffusion between communities, whether via genetic influence or learning.

3. For termite fishing development, Gombe: Goodall (1968a: 208), Lonsdorf *et al.* (2004), and Lonsdorf (2005, 2006). Interestingly, McGrew (1977) noted at Gombe that ant dipping development differed from that of termite fishing, probably because ants sting. Infants have comparatively few opportunities to observe their mothers and tend to watch while clinging or from a distance. As they mature, they practice little and do not achieve proficiency until adolescence. Goualougo: Musgrave *et al.* (2016), who consider tool transfers between mothers and offspring a form of "teaching," based on a functional definition of teaching that does not include mental state attribution by instructors, which is a characteristic of teaching in humans (Caro and Hauser, 1992).

   For nut-cracking development, Bossou: Matsuzawa (1994), Inoue-Nakamura and Matsuzawa (1997), and Biro *et al.* (2003, 2010a). Matsuzawa (1994) also speculated that there is a critical learning period for the behavior, since one individual who was encumbered for several months by a snare as a youngster failed to become proficient later on after the snare was removed; Taï N: Boesch (1991c), who also considered tool transfers a form of teaching.

4. Whiten *et al.* (2005). See reviews of additional diffusion experiments in Whiten (2010) and Whiten *et al.* (2016). See Tomasello and Call (2010) for a review of captive research on chimpanzee social cognition, which is integral to investigating mechanisms of social learning.

5. Whiten *et al.* (2007).

6. Community variation in self-medicative, whole-leaf swallowing was first described by Wrangham and Nishida (1983) for Gombe and Mahale. Subsequent studies are reviewed in Huffman (2015). Huffman and Hirata (2004) and Huffman *et al.* (2010) conducted experimental studies that point to both predisposition for and social influences on the acquisition of the behavior.

7. Gombe: O'Malley *et al.* (2012) documented the spread of "fishing" for a new species of ant beginning in 1992, a behavior possibly introduced by female immigrants from a neighboring community and now customary in the community; Budongo: Hobaiter *et al.* (2014b) documented the invention and diffusion of a previously unobserved moss-sponging behavior over a six-day period in 2011. The behavior was still present in the community three years later and had spread to additional individuals (Lamon *et al.*, 2017). For examples of short-term diffusion of novel behaviors, Gombe: Goodall (1973: 166); Mahale: Nishida *et al.* (2009); Bossou: Ohashi and Matsuzawa (2011); Taï N: Boesch (1995).

8. In a review of army ant predation data from fourteen field sites, Schöning *et al.* (2008) concluded that some intersite variation could be explained by differences

in prey characteristics. Similarly, Sanz *et al.* (2014) found that some of the variation between the tools used to harvest termites by Goualougo (Republic of the Congo) and La Belgique (Cameroon) chimpanzees was related to termite nest structure. In contrast, based on an experimental study in which researchers simulated ant fishing behavior, Möbius *et al.* (2008) found that differences in tool length between Bossou and Taï NS could not be explained by differences in ant behavior between the two sites. Similarly, Luncz *et al.* (2012) and Luncz and Boesch (2015) were unable to attribute differences in nut-cracking and insect foraging techniques between neighboring Taï NSE communities to microecological factors, and Koops *et al.* (2015) found that variation between neighboring Kalinzu Forest communities in the length of tools used to harvest army ants was not related to the availability of prey species.

9. Goodall (1970: 195, 1986: 536).

10. Bentley-Condit and Smith (2010) review the literature on animal tool use, organizing reported observations into ten categories: food preparation, food extraction, food transport, food capture, physical maintenance, mate attraction, nest construction, predator defense, agonism, and other.

11. Detailed descriptions of tool-use behaviors at Gombe (Goodall, 1968a: 202–208, 1970, 1986: 535–564) and Mahale (Nishie, 2015) provide a broad overview of the range and flexibility of chimpanzee tool use.

12. For reviews of a variety of models developed to evaluate tool complexity and its cognitive significance, see Matsuzawa (1996), Visalberghi and Fragaszy (2006, for tool use in capuchin monkeys), Sanz and Morgan (2010), Humle and Fragaszy (2011), and Boesch (2013). See Carvalho *et al.* (2008) for descriptions of Bossou chimpanzees using wedges to stabilize anvils during nut cracking. See Sanz and Morgan (2009), Boesch *et al.* (2009), and Estienne *et al.* (2017) for analyses of sequential tool use to extract honey at Goualougo (Republic of Cameroon) and Loango (Gabon), and for reviews of honey gathering in different communities.

13. Nakamura (2002: Table 5.1) lists additional (possible) community-level variants in the social domain.

14. See Whiten *et al.* (2001) for detailed descriptions of specific behavior patterns.

15. Most studies of time spent grooming do not distinguish between grooming types. Where they have, the following percentages have been reported for mutual grooming. Gombe: 15 and 22 percent of male–male and male–female grooming dyads, respectively, during greetings included mutual grooming (estimated from Goodall 1986: Fig. 14.4); Mahale: 12.6 percent of grooming time for alpha male–other male dyads (calculated from Kawanaka, 1990: Table 8.3); Taï N: 23 and 40 percent of male–male and male–female grooming time, respectively, in 1993 (Boesch and Boesch-Achermann, 2000: Table 6.8, which also reports inexplicably high values of 73 and 60 percent, respectively, for 1988); Taï S: 25 percent of pooled adult grooming time between 2003 and 2006 (Gomes *et al.*, 2009); Kanyawara: 11 percent of adult male–adult male grooming time, in which 28 percent of bouts included at least some mutual grooming. Of those, 27 percent of the bout was mutual (Machanda *et al.*, 2014).

16. "Branch-clasp," first described by Goodall (1965: 470), was customary or habitual in all nine communities compared by Whiten *et al.* (1999).

17. McGrew and Tutin (1978: 238). "Hand-clasp" grooming was first identified as a community-distinctive behavior when these two Gombe researchers, who had never seen the behavior in their own study population, documented its repeated occurrence at Mahale during a brief visit there. See Nakamura

(2002) and McGrew (2004: 139) for discussions of the relationship between branch- and hand-clasp grooming.

18. Kanyawara: In a study of different forms of high-arm grooming, Wrangham *et al.* (2016) found that the duration high-arm grooming periods ranged from 3 to 155 seconds.

19. In the initial survey, hand-clasp grooming was customary or habitual in five communities (Mahale K and M, Taï N, Kanyawara, and Lopé, Gabon) and absent in three (Gombe, Bossou, and Budongo), and its status not yet established in one (Mt. Assirik, Senegal) (Whiten *et al.*, 2001: Table 3). It has since been reported as customary or habitual at four more sites: Ngogo (J. C. Mitani and D. P. Watts, personal communication in Nakamura, 2002: 75), Fongoli in Senegal (J. D. Pruetz, personal communication in Humle *et al.*, 2009b; W. C. McGrew, personal communication in Webster *et al.*, 2009); Goualougo in Republic of the Congo (W. C. McGrew, personal communication in Webster *et al.*, 2009); and Kalinzu in Uganda (C. Hashimoto and T. Furuichi, personal communication in Nakamura, 2002: 75). At Taï N, where it was reported to occur habitually in Whiten *et al.* (1999), the behavior appeared and was documented for two years and then disappeared (C. Boesch, personal communication in Humle *et al.*, 2009b). A single occurrence has been reported at Bossou (G. Yamakoshi, personal communication in Humle *et al.*, 2009b).

20. In a study of "grooming hand-clasp" behavior at Gombe and Mahale, McGrew *et al.* (2001: 150) distinguished palm-to-palm from non-palm-to-palm, which "usually involves flexed wrists, and one limb usually rests on the other, which bears at least some of the weight of both." Wrangham *et al.* (2016: 3033) proposed substituting the term "high-arm mutual grooming" for "hand-clasp grooming" to include "episodes of mutual grooming when two partners sit facing each other, each holding an elbow higher than either of their shoulders, and their raised hand, wrist, or arm is touching the similarly raised hand, wrist, or arm of the partner."

21. For example, percentage of high-arm grooming bouts that were "palm-to-palm clasp" at Mahale: 41.7 percent in K-group, 5.3 percent in M-group (Nakamura and Uehara, 2004); at Kanyawara: 32.6 ± 18.9 percent overall (mean $\pm$ standard deviation), varying across individuals from 6.8 to 57.8 percent (Wrangham *et al.*, 2016).

22. Gombe: Goodall (1968a: 202; 1986: 391).

23. Mahale: Zamma (2002).

24. Mahale: Nishida (1980); Nakamura *et al.* (2000).

25. Gombe: Shimada (2002); Ngogo: Nishida *et al.* (2004), Nakamura (2010).

26. Various researchers from the 1950s onward called group-level behavioral variation in primate species "sub-cultural," "pre-cultural," "proto-cultural," or just "cultural" variation. Widespread reference to chimpanzee "culture," both by researchers and in popular writing, followed the publication of two books in particular, *Chimpanzee Material Culture* (McGrew, 1992) and *Chimpanzee Cultures* (Wrangham *et al.*, 1994).

27. McGrew and Tutin (1978) and McGrew (1992, 2004), based on Kroeber (1928). It is noteworthy that Kroeber considered apes cultureless, for want of language, and some human societies as nearly so: "By common consent the study of the earliest and most backward peoples and cultures has been left to Anthropology. The significance of these incipient cultures is obviously not so much intrinsic as in the light which they may shed by enabling a wider and fuller range of comparisons on the nature of culture as a whole" (p. 325).

28. Three possible learning mechanisms are typically invoked: (1) "local enhancement" or "stimulus enhancement," "in which one individual draws the attention of another to a locale or to a stimulus and, thus, increases the probability of successful discovery and learning by the naive individual"; (2) "imitative learning," "in which the learner actually copies another's behavior"; and (3) "emulation learning," in which an individual attempts "to reproduce the observed end result … without copying the behavioral methods of the demonstrator" (Tomasello, 1994: 303–304). See also reviews of ape learning research in Tomasello and Call (2010) and Whiten (2012, 2015).

29. A wide range of theoretical perspectives concerning the appropriateness of applying the concept of culture to animals is presented in Fragaszy and Perry (2003a) and Laland and Galef (2009), edited volumes with contributions from the fields of zoology, anthropology, psychology, and philosophy. Fragaszy (2003) and Fragaszy and Perry (2003b) offer grounds for using the more cautious term "traditions" to describe learned group-level behavioral variation in nonhuman animals, since "culture" in their view encompasses some special human attributes.

30. Tomasello (1994, 2009) identifies three unique characteristics of human cultural traditions that are a direct consequence of distinctive human learning processes: (1) "universality" (virtually all group members practice the behavior), (2) "uniformity" (a high degree of similarity in performance between individuals), and (3) the so-called "ratchet effect," or the "accumulation of modifications over generations" (1994: 312). See also Tennie *et al.* (2009b) for a more recent explication of the ratchet effect, which in their view relies as well on "the facts that (i) human social learning is more oriented towards process than product and (ii) unique forms of human cooperation lead to active teaching, social motivations for conformity and normative sanctions against non-conformity" (p. 2405).

31. Hill (2009) offers a persuasive argument for restricting the term "culture" to humans, based on the unique kinds of information transmitted both vertically and horizontally in human groups.

32. See Reader and Laland (2002) and Reader *et al.* (2011) for analyses that demonstrate the correlation between comparative brain size, which is high in chimpanzees, and the propensity for innovation in primates. Whiten *et al.* (2001: 1483, 1486, 1506) conclude that intergroup behavioral variation in chimpanzees is exceeded only in humans.

33. Whiten *et al.* (2005, 2007).

34. Luncz and Boesch (2014).

35. See van Leeuwen and Haun (2014) for a general critique of conformity interpretations of experimental results from captive studies; Ngamba Island Chimpanzee Sanctuary (Uganda): Tennie *et al.* (2012) found that individuals were resistant to imitating novel behaviors to receive food rewards; Chimfunshi Wildlife Orphanage (Zambia): van Leeuwen *et al.* (2013) found that individuals abandoned their first-learned strategy in token-for-food experiments to maximize payoffs.

36. Mahale: Nakamura and Nishida (2013) observed mothers actively molding the grooming posture of their offspring; Kanyawara: Wrangham *et al.* (2016) found that frequencies of palm-to-palm high-arm grooming were consistent within matrilines; Budongo: The moss-sponging behavior described previously (endnote 7) appears to have initially spread horizontally when individuals observed nearby group mates performing the behavior but subsequently radiated through matrilines (Lamon *et al.*, 2017); Yerkes (Georgia,

USA): Bonnie and de Waal (2006) found that the formation of high-arm grooming partnerships was correlated with rates of grooming and proximity between participants; Chimfunshi Wildlife Orphanage (Zambia): van Leeuwen *et al.* (2012) found that the first partner of new high-arm groomers was usually the mother; Cronin *et al.* (2014) found that variation in grouping tendencies between groups reflected the effects of key individuals within groups that could monopolize provisioned resources; van Leeuwen *et al.* (2017), in a follow-up of their (2012) study, found that the transmission of high-arm grooming patterns also extended beyond matrilines within groups, perhaps influenced by regular, artificial social aggregations resulting from provisioning.

37. See Perry (2009: 266) for a convincing discussion of why linking socially learned behavioral traits to group identity is probably absent in nonhuman primates, since unlike humans they do not form cooperative relationships with strangers.

EPILOGUE

1. Goodall (1971: 251).

APPENDIX: FIELD METHODS FOR STUDYING WILD CHIMPANZEES

1. Full habituation could take as long as ten years, especially for females (M. Muller, personal communication). Note also that it possible that more cautious individuals learn to be less fearful by monitoring the behavior of more intrepid group mates (Samuni *et al.*, 2014).

2. This appendix only describes a fraction of the sampling methods available to animal behavior researchers. Comprehensive accounts may be found in Altmann (1974) and Martin and Bateson (1995), as well as in most animal behavior textbooks.

3. Nishida (2010).

References

Allen, M. (1981). Individual copulatory preference and the "strange female effect" in a captive group-living male chimpanzee (*Pan troglodytes*). *Primates* 22: 221–236.

Altmann, J. (1974). Observational study of behavior: Sampling methods. *Behaviour* 49: 227–267.

Amsler, S. J. (2010). Energetic costs of territorial boundary patrols by wild chimpanzees. *American Journal of Primatology* 72: 93–103.

Amster, G. and Sella, G. (2016). Life history effects on the molecular clock of autosomes and sex chromosomes. *Proceedings of the National Academy of Sciences of the United States of America* 113: 1588–1593.

Anderson, D. P., Nordheim, E. V., Boesch, C., and Moermond, T. C. (2002). Factors influencing fission-fusion grouping in chimpanzees in the Taï National Park, Cote d'Ivoire. In *Behavioural Diversity in Chimpanzees and Bonobos* (C. Boesch, G. Hohmann, and L. F. Marchant, eds.), Cambridge University Press, Cambridge, pp. 90–101.

Anderson, D. P., Nordheim, E. V., Moermond, T. C., Gone Bi, Z. B., and Boesch, C. (2005). Factors influencing tree phenology in Taï National Park, Côte d'Ivoire. *Biotropica* 37: 631–640.

Andrews, K. (2017). Chimpanzee mind reading: Don't stop believing. *Philosophy Compass* 12: e12394. doi:10.1111/phc3.12394

Arcadi, A. Clark (1996). Phrase structure of wild chimpanzee pant hoots: Patterns of production and interpopulation variability. *American Journal of Primatology* 39: 159–178.

Arcadi, A. Clark (2000). Vocal responsiveness in male wild chimpanzees: Implications for the evolution of language. *Journal of Human Evolution* 39: 205–223.

Arcadi, A. Clark (2005). Language evolution: What do chimpanzees have to say? *Current Biology* 15: R884–R886.

Arcadi, A. Clark and Wallauer, W. (2011). Individual-level lateralization in the asymmetrical gaits of wild chimpanzees (*Pan troglodytes*): Implications for hand preference and skeletal asymmetry? *Behaviour* 148: 1419–1441.

Arcadi, A. Clark and Wallauer, W. (2013). They wallop like they gallop: Audiovisual analysis reveals the influence of gait on buttress drumming by wild chimpanzees (*Pan troglodytes*). *International Journal of Primatology* 34: 194–215.

Arcadi, A. Clark and Wrangham, R. W. (1999). Infanticide in chimpanzees: Review of cases and a new within-group observation from the Kanyawara study group in Kibale National Park. *Primates* 40(2): 337–351.

Arcadi, A. Clark, Robert, D., and Boesch, C. (1998). Buttress drumming by wild chimpanzees: Temporal patterning, phrase integration into loud calls, and preliminary evidence for individual distinctiveness. *Primates* 39(4): 505–518.

Arcadi, A. Clark, Robert, D., and Mugurusi, F. (2004). A comparison of buttress drumming by male chimpanzees from two populations. *Primates* 45(2): 135–139.

Arnold, K. and Whiten, A. (2001). Post-conflict behaviour of wild chimpanzees (*Pan troglodytes schweinfurthii*) in the Budongo Forest, Uganda. *Behaviour* 138: 649–690.

Arnold, K. and Whiten, A. (2003). Grooming interactions among the chimpanzees of the Budongo Forest, Uganda: Test of five explanatory models. *Behaviour* 140: 519–552.

Arnold, K., Fraser, O. N., and Aureli, F. (2011). Postconflict reconciliation. In *Primates in Perspective* (C. J. Campbell, A. Fuentes, K. C. MacKinnon, S. K. Bearder, and R. M. Stumpf, eds.), 2nd edn., Oxford University Press, Oxford, pp. 608–625.

Atsalis, S. and Videan, E. (2009). Reproductive aging in captive and wild common chimpanzees: Factors influencing the rate of follicular depletion. *American Journal of Primatology* 71: 271–282.

Aureli, F. and de Waal, F. B. M., eds. (2000). *Natural Conflict Resolution*. University of California Press, Berkeley, CA.

Aureli, F. and Smucny, D. (2000). The role of emotion in conflict and conflict resolution. In *Natural Conflict Resolution* (F. Aureli and F. B. M. de Waal, eds.), University of California Press, Berkeley, CA, pp. 199–224.

Aureli, F., Preston, S. D., and de Waal, F. B. M. (1999). Heart rate responses to social interactions in free-moving rhesus macaques (*Macaca mulatta*): A pilot study. *Journal of Comparative Psychology* 113: 59–65.

Aureli, F., Cords, M., and van Schaik, C. P. (2002). Conflict resolution following aggression in gregarious animals: A predictive framework. *Animal Behaviour* 64: 325–343.

Aureli, F., Schaffner, C. M., Boesch, C., Bearder, S. K., Call, J., Chapman, C. A., Connor, R., Di Fiore, A., Dunbar, R. I. M., Henzi, S. P., Holekamp, K., Korstjens, A. H., Layton, R., Lee, P., Lehmann, J., Manson, J. H., Ramos-Fernandez, G., Strier, K. B., and van Schaik, C. P. (2008). Fission-fusion dynamics. *Current Anthropology* 49: 627–654.

Aureli, F., Fraser, O. N., Schaffner, C. M., and Schino, G. (2012). The regulation of social relationships. In *The Evolution of Primate Societies* (J. C. Mitani, J. Call, P. M. Kappeler, R. A. Palombit, and J. B. Silk, eds.), University of Chicago Press, Chicago, IL, pp. 531–551.

Babiszewska, M., Schel, A. M., Wilke, C., and Slocombe, K. E. (2015). Social, contextual, and individual factors affecting the occurrence and acoustic structure of drumming bouts in wild chimpanzees (*Pan troglodytes*). *American Journal of Physical Anthropology* 156: 125–134.

Babweteera, F., Sheil, D., Reynolds, V., Plumptre, A.J., Zuberbühler, K., Hill, C.M., Webber, A., and Tweheyo, M. (2011). Environmental and anthropogenic changes in and around Budongo Forest Reserve. In *The Ecological Impact of Long-Term Changes in Africa's Rift Valley* (A. J. Plumptre, ed.), Nova Science, New York, NY, pp. 31–53.

Balcomb, S. R., Chapman, C. A., and Wrangham, R. W. (2000). Relationship between chimpanzee (*Pan troglodytes*) density and large, fleshy-fruit tree density: Conservation implications. *American Journal of Primatology* 51: 197–203.

Baldwin, L. A. and Teleki, G. (1973). Field research on chimpanzees and gorillas: An historical, geographical, and bibliographical listing. *Primates* 14: 315–330.

Ban, S. D., Boesch, C., and Janmaat, K. R. L. (2014). Taï chimpanzees anticipate revisiting high-valued fruit trees from further distances. *Animal Cognition* 17: 1353–1364.

Ban, S. D., Boesch, C., N'Guessan, A., N'Goran, E. K., Tako, A., and Janmaat, K. R. L. (2016). Taï chimpanzees change their travel direction for rare feeding trees providing fatty fruits. *Animal Behaviour* 118: 135–147.

Bartlett, T. Q. (2011). The Hylobatidae: Small apes of Asia. In *Primates in Perspective* (C. J. Campbell, A. Fuentes, K. C. MacKinnon, S. K. Bearder, and R. M. Stumpf, eds.), 2nd edn., Oxford University Press, Oxford, pp. 300–312.

Bates, L. A. and Byrne, R. W. (2009). Sex differences in the movement patterns of free-ranging chimpanzees (*Pan troglodytes schweinfurthii*): Foraging and border checking. *Behavioral Ecology and Sociobiology* 64: 247–255.

Bentley-Condit, V. K. and Smith, E. O. (2010). Animal tool use: Current definitions and an updated comprehensive catalog. *Behaviour* 147: 185–221.

Biro, D., Inoue-Nakamura, N., Tonooka, R., Yamakoshi, G., Sousa, C., and Matsuzawa, T. (2003). Cultural innovation and transmission of tool use in wild chimpanzees: Evidence from field experiments. *Animal Cognition* 6: 213–223.

Biro, D., Carvalho, S., and Matsuzawa, T. (2010a). Tools, traditions, and technologies: Interdisciplinary approaches to chimpanzee nut cracking. In *The Mind of the Chimpanzee* (E. V. Lonsdorf, S. R. Ross, and T. Matsuzawa, eds.), University of Chicago Press, Chicago, IL, pp. 141–155.

Biro, D., Humle, T., Koops, K., Sousa, C., Hayashi, M., and Matsuzawa, T. (2010b). Chimpanzee mothers at Bossou, Guinea carry the mummified remains of their dead infants. *Current Biology* 20: R351–R352.

Bjork, A., Liu, W., Wertheim, J. O., Hahn, B. H., and Worobey, M. (2011). Evolutionary history of chimpanzees inferred from complete mitochondrial genomes. *Molecular Biology and Evolution* 28: 615–623.

Boesch, C. (1991a). The effects of leopard predation on grouping patterns in forest chimpanzees. *Behaviour* 117: 220–241.

Boesch, C. (1991b). Symbolic communication in wild chimpanzees? *Human Evolution* 6: 81–90.

Boesch, C. (1991c). Teaching among wild chimpanzees. *Animal Behaviour* 41: 530–532.

Boesch, C. (1994). Cooperative hunting in wild chimpanzees. *Animal Behaviour* 48: 653–667.

Boesch, C. (1995). Innovation in wild chimpanzees (*Pan troglodytes*). *International Journal of Primatology* 16: 1–16.

Boesch, C. (1996). Social grouping in Taï chimpanzees. In *Great Ape Societies* (W. C. McGrew, L. F. Marchant, and T. Nishida, eds.), Cambridge University Press, Cambridge, pp. 101–113.

Boesch, C. (2008). Why do chimpanzees die in the forest? The challenges of understanding and controlling for wild ape health. *American Journal of Primatology* 70: 722–726.

Boesch, C. (2013). Ecology and cognition of tool use in chimpanzees. In *Tool Use in Animals: Cognition and Ecology* (C. M. Sanz, J. Call, and C. Boesch, eds.), Cambridge University Press, Cambridge, pp. 21–47.

Boesch, C. and Boesch, H. (1983). Optimization of nut-cracking with natural hammers by wild chimpanzees. *Behaviour* 83(3–4): 265–285.

Boesch, C. and Boesch, H. (1989). Hunting behavior of wild chimpanzees in the Taï National Park. *American Journal of Physical Anthropology* 78: 547–573.

Boesch, C. and Boesch-Achermann, H. (2000). *The Chimpanzees of the Taï Forest.* Oxford University Press, Oxford.

Boesch, C., Hohmann, G., and Marchant, L. F., eds. (2002). *Behavioural Diversity in Chimpanzees and Bonobos.* Cambridge University Press, Cambridge.

Boesch, C., Kohou, G., Néné, H., and Vigilant, L. (2006). Male competition and paternity in wild chimpanzees of the Taï Forest. *American Journal of Physical Anthropology* 130: 108–115.

Boesch, C., Crockford, C., Herbinger, I., Wittig, R., Moebius, Y., and Normand, E. (2008). Intergroup conflicts among chimpanzees in Taï National Park: Lethal violence and the female perspective. *American Journal of Primatology* 70: 519–532.

Boesch, C., Head, J., and Robbins, M. M. (2009). Complex tool sets for honey extraction among chimpanzees in Loango National Park, Gabon. *Journal of Human Evolution* 56: 560–569.

Boesch, C., Bolé, C., Eckhardt, N., and Boesch, H. (2010). Altruism in forest chimpanzees: The case of adoption. *PLoS One* 5(1): e8901.

Bonnie, K. E. and de Waal, F. B. M. (2006). Affiliation promotes the transmission of a social custom: Handclasp grooming among captive chimpanzees. *Primates* 47: 27–34.

Bradbury, J. W. and Vehrencamp, S. L. (2011). *Principles of Animal Communication.* Sinauer Associates, Sunderland, MA.

Brand, C., Eguma, R., Zuberbühler, K., and Hobaiter, C. (2014). First report of prey capture from human laid snare-traps by wild chimpanzees. *Primates* 55: 437–440.

Busse, C. D. (1977). Chimpanzee predation as a possible factor in the evolution of red colobus monkey social organization. *Evolution* 31: 907–911.

Busse, C. D. (1978). Do chimpanzees hunt cooperatively? *American Naturalist* 112: 767–770.

Butynski, T. M. (2003). The robust chimpanzee *Pan troglodytes*: Taxonomy, distribution, abundance, and conservation status. In *West African Chimpanzees: Status Survey and Conservation Action Plan* (R. Kormos, C. Boesch, M. I. Bakarr, and T. Butynski, eds.), IUCN/SSC Primate Specialist Group, IUCN, Gland, Switzerland, pp. 5–12.

Bygott, J. D. (1979). Agonistic behavior, dominance, and social structure in wild chimpanzees of the Gombe National Park. In *The Great Apes* (D. A. Hamburg and E. R. McCown, eds.), Benjamin/Cummings, Menlo Park, CA, pp. 405–427.

Call, J. and Tomasello, M. (2007). The gestural repertoire of chimpanzees (*Pan troglodytes*). In *The Gestural Communication of Apes and Monkeys* (J. Call and M. Tomasello, eds.), Lawrence Erlbaum Associates, Mahwah, NJ, pp. 17–39.

Carbone, L. *et al.* (2014). Gibbon genome and the fast karyotype evolution of small apes. *Nature* 513: 195–201.

Caro, T. M. and Hauser, M. D. (1992). Is there teaching in animals? *Quarterly Review of Biology* 67: 151–174.

Carvalho, S., Cunha, E., Sousa, C., and Matsuzawa, T. (2008). Chaînes opératoires and resource-exploitation strategies in chimpanzee (*Pan troglodytes*) nut cracking. *Journal of Human Evolution* 55: 148–163.

Chapman, C. A. and Wrangham, R. W. (1993). Range use of the forest chimpanzees of Kibale: Implications for the understanding of chimpanzee social organization. *American Journal of Primatology* 31: 263–273.

Chapman, C. A., White, F. J., and Wrangham, R. W. (1994). Party size in chimpanzees and bonobos. In *Chimpanzee Cultures* (R. W. Wrangham, W. C. McGrew, F. B. M. de Waal, and P. Heltne, eds.), Harvard University Press, Cambridge, MA, pp. 41–57.

Chapman, C. A., Wrangham, R. W., Chapman, L. J., Kennard, D. K., and Zanne, A. E. (1999). Fruit and flower phenology at two sites in Kibale National Park, Uganda. *Journal of Tropical Ecology* 15: 189–211.

Chapman, C. A., Chapman, L. J., Struhsaker, T. T., Zanne, A. E., Clark, C. J., and Poulsen, J. R. (2005). A long-term evaluation of fruiting phenology: Importance of climate change. *Journal of Tropical Ecology* 21: 31–45.

Chapman, C. A., Chapman, L. J., Ghai, R., Hartter, J., Jacob, A. L., Lwanga, J. S., Omeja, P., Rothman, J. M., and Twinomugisha, D. (2011). Complex responses to climate and anthropogenic changes: An evaluation based on long-term data from Kibale National Park, Uganda. In *The Ecological Impact of Long-term Changes in Africa's Rift Valley* (A. J. Plumptre, ed.), Nova Science, New York, NY, pp. 73–94.

Cheney, D. L. and Seyfarth, R. M. (2007). *Baboon Metaphysics.* University of Chicago Press, Chicago, IL.

Clark (Arcadi), A. P. (1993). Rank differences in the production of vocalizations by wild chimpanzees as a function of social context. *American Journal of Primatology* 31: 159–179.

Clark (Arcadi), A. P. and Wrangham, R. W. (1993). Acoustic analysis of wild chimpanzee pant hoots: Do Kibale Forest chimpanzees have an acoustically distinct food arrival pant hoot? *American Journal of Primatology* 31: 99–109.

Clark (Arcadi), A. P. and Wrangham, R. W. (1994). Chimpanzee arrival pant hoots: Do they signify food or status? *International Journal of Primatology* 15: 185–205.

Clark, C. B. (1977). A preliminary report on weaning among chimpanzees of the Gombe National Park, Tanzania. In *Primate Bio-social Development* (S. Chevalier-Skolnikoff and F. E. Poirier, eds.), Garland Publishing, New York, NY, pp. 235–260.

Connor, R. (2007). Dolphin social intelligence: Complex alliance relationships in bottlenose dolphins and a consideration of selective environments for extreme brain size evolution in mammals. *Philosophical Transactions of the Royal Society B* 362: 587–602.

Constable, J. L., Ashley, M. V., Goodall, J., and Pusey, A. E. (2001). Noninvasive paternity assignment in Gombe chimpanzees. *Molecular Ecology* 10: 1279–1300.

Crockford, C. and Boesch, C. (2003). Context-specific calls in wild chimpanzees, *Pan troglodytes verus*: Analysis of barks. *Animal Behaviour* 66: 115–125.

Crockford, C. and Boesch, C. (2005). Call combinations in wild chimpanzees. *Behaviour* 142: 397–421.

Crockford, C., Herbinger, I., Vigilant, L., and Boesch, C. (2004). Wild chimpanzees produce group-specific calls: A case for vocal learning? *Ethology* 110: 221–243.

Crockford, C., Wittig, R. M., Whitten, P. L., Seyfarth, R. M., and Cheney, D. L. (2007). Baboons eavesdrop to deduce mating opportunities. *Animal Behaviour* 73: 885–890.

Crockford, C., Wittig, R. M., Whitten, P. L., Seyfarth, R. M., and Cheney, D. L. (2008). Social stressors and coping mechanisms in wild female baboons (*Papio hamadryas ursinus*). *Hormones and Behavior* 53: 254–265.

Crockford, C., Wittig, R., and Zuberbühler, K. (2012). Chimpanzees distinguish acoustically similar alert hoos from resting hoos. *Folia Primatologica* 84: 260–261.

Crockford, C., Wittig, R. M., Langergraber, K., Ziegler, T.E., Zuberbühler, K., and Deschner, T. (2013). Urinary oxytocin and social bonding in related and unrelated wild chimpanzees. *Proceedings of the Royal Society B* 280: 1–8. http://dx.doi.org/10.1098/rspb.2012.2765

Cronin, K. A., van Leeuwen, E. J. C., Vreeman, V., and Haun, D. B. M. (2014). Population-level variability in the social climates of four chimpanzee societies. *Evolution and Human Behavior* 35: 389–396.

Darwin, C. (1871). *The Descent of Man, and Selection in Relation to Sex.* Princeton University Press, Princeton, NJ.

Davila-Ross, M., Allcock, B., Thomas, C., and Bard, K. A. (2011). Aping expressions? Chimpanzees produce distinct laugh types when responding to laughter of others. *Emotion* 11: 1013–1020.

Davila-Ross, M., Jesus, G., Osborne, J., and Bard, K. A. (2015). Chimpanzees (*Pan troglodytes*) produce same types of "laugh faces" when they emit laughter and when they are silent. *PLoS ONE* 10(6): e0127337. doi:10.1371/journal.pone.0127337

de Waal, F. B. M. (1982/1998). *Chimpanzee Politics.* Johns Hopkins University Press, Baltimore, MD.

de Waal, F. B. M. (1994). Chimpanzee's adaptive potential: A comparison of social life under captive and wild conditions. In *Chimpanzee Cultures* (R. W. Wrangham, W. C. McGrew, F. B. M. de Waal, and P. Heltne, eds.), Harvard University Press, Cambridge, MA, pp. 243–260.

de Waal, F. B. M. and Aureli, F. (1996). Consolation, reconciliation, and a possible cognitive difference between macaques and chimpanzees. In *Reaching into Thought: The Minds of the Great Apes* (A. E. Russon, K. A. Bard, and S. T. Parker, eds.), Cambridge University Press, Cambridge, pp. 80–110.

de Waal, F. B. M. and Tyack, P. L., eds. (2003). *Animal Social Complexity.* Harvard University Press, Cambridge, MA.

de Waal, F. B. M. and van Roosmalen, A. (1979). Reconciliation and consolation among chimpanzees. *Behavioral Ecology and Sociobiology* 5: 55–66.

de Waal, F. B. M. and Yoshihara, D. (1983). Reconciliation and redirected affection in rhesus monkeys. *Behaviour* 85: 224–241.

Dennett, D. C. (1987). *The Intentional Stance.* Massachusetts Institute of Technology Press, Cambridge, MA.

Deschner, T., Heistermann, M., Hodges, K., and Boesch, C. (2003). Timing and probability of ovulation in relation to sex skin swelling in wild West African chimpanzees, *Pan troglodytes verus. Animal Behaviour* 66: 551–560.

Deschner, T., Heistermann, M., Hodges, K., and Boesch, C. (2004). Female sexual swelling size, timing of ovulation, and male behavior in wild West African chimpanzees. *Hormones and Behavior* 46: 204–215.

Dominy, N. J., Svenning, J., and Li, W. (2003). Historical contingency in the evolution of primate color vision. *Journal of Human Evolution* 44: 25–45.

Doran, D. (1997). Influence of seasonality on activity patterns, feeding behavior, ranging, and grouping patterns in Taï chimpanzees. *International Journal of Primatology* 18: 183–206.

Duffy, K. G., Wrangham, R. W., and Silk, J. B. (2007). Male chimpanzees exchange political support for mating opportunities. *Current Biology* 17: R586.

Dunbar, R. I. M. (1988). *Primate Social Systems.* Cornell University Press, Ithaca, NY.

Dunbar, R. I. M. (1998). The social brain hypothesis. *Evolutionary Anthropology* 6: 178–190.

Dunbar, R. I. M. and Shultz, S. (2007). Understanding primate brain evolution. *Philosophical Transactions of the Royal Society B* 362: 649–658.

Emery Thompson, M. (2005). Reproductive endocrinology of wild female chimpanzees (*Pan troglodytes troglodytes*): Methodological considerations and the role of hormones in sex and conception. *American Journal of Primatology* 67: 137–158.

Emery Thompson, M. (2013). Reproductive ecology of female chimpanzees. *American Journal of Primatology* 75: 222–237.

Emery Thompson, M. and Wrangham, R. W. (2006). Comparison of sex differences in gregariousness in fission-fusion species. In *Primates of Western Uganda* (N. E. Newton-Fisher, H. Notman, J. D. Paterson, and V. Reynolds, eds.), Springer, New York, NY, pp. 209–226.

Emery Thompson, M. and Wrangham, R. W. (2008). Diet and reproductive function in wild female chimpanzees (*Pan troglodytes schweinfurthii*) at Kibale National Park, Uganda. *American Journal of Physical Anthropology* 135: 171–181.

Emery Thompson, M., Newton-Fisher, N. E., and Reynolds, V. (2006). Probable community transfer of parous adult female chimpanzees in the Budongo Forest, Uganda. *International Journal of Primatology* 27: 1601–1617.

Emery Thompson, M., Kahlenberg, S. M., Gilby, I. C., and Wrangham, R. W. (2007a). Core area quality is associated with variance in reproductive success among female chimpanzees at Kibale National Park. *Animal Behaviour* 73: 501–512.

Emery Thompson, M., Jones, J. H., Pusey, A. E., Brewer-Marsden, S., Goodall, J., Marsden, D., Matsuzawa, T., Nishida, T., Reynolds, V., Sugiyama, Y., and Wrangham, R. W. (2007b). Age and fertility patterns in wild chimpanzees provide insights into the evolution of menopause. *Current Biology* 17: 1–7.

Emery Thompson, M., Muller, M. N., and Wrangham, R. W. (2012). The energetics of lactation and the return to fecundity in wild chimpanzees. *Behavioral Ecology* 23(6): 1234–1241.

Emery Thompson, M., Muller, M. N., and Wrangham, R. W. (2014). Male chimpanzees compromise the foraging success of their mates in Kibale National Park, Uganda. *Behavioral Ecology and Sociobiology* 68: 1973–1983.

Estienne, V., Stephens, C., and Boesch, C. (2017). Extraction of honey from underground bee nests by central African chimpanzees (*Pan troglodytes troglodytes*) in Loango National Park, Gabon: Techniques and individual differences. *American Journal of Primatology.* doi:10.1002/ajp.22672

Fallon, B.L., Neumann, C., Byrne, R. W., and Zuberbühler, K. (2016). Female chimpanzees adjust copulation calls according to reproductive status and level of female competition. *Animal Behaviour* 113: 87–92.

Fedurek, P. and Slocombe, K. E. (2013). The social function of food-associated calls in male chimpanzees. *American Journal of Primatology* 75: 726–739.

Fedurek, P., Donnellan, E., and Slocombe, K. E. (2014). Social and ecological correlates of long-distance pant hoot calls in male chimpanzees. *Behavioral Ecology and Sociobiology* 68: 1345–1355.

Fedurek, P., Schell, A. M., and Slocombe, K. E. (2013a). The acoustic structure of chimpanzee pant-hooting facilitates chorusing. *Behavioral Ecology and Sociobiology* 67: 1781–1789.

Fedurek, P., Machanda, Z. P., Schel, A. M., and Slocombe, K. E. (2013b). Pant hoot chorusing and social bonds in male chimpanzees. *Animal Behaviour* 86: 189–196.

Fedurek, P., Slocombe, K. E., and Zuberbühler, K. (2015a). Chimpanzees communicate to two different audiences during aggressive interactions. *Animal Behaviour* 110: 21–28.

Fedurek, P., Slocombe, K.E., Hartel, J.A., and Zuberbühler, K. (2015b). Chimpanzee lip-smacking facilitates cooperative behaviour. *Scientific Reports* 5: 13460. doi:10.1038/srep13460

Fedurek, P., Slocombe, K. E., Enigk, D. K., Emery Thompson, M., Wrangham, R. W., and Muller, M. N. (2016). The relationship between testosterone and long-distance calling in wild male chimpanzees. *Behavioral Ecology and Sociobiology* 70: 659–672.

Feldblum, J. T., Wroblewski, E. E., Rudicell, R. S., Hahn, B. H., Paiva, T., Cetinkaya-Rundel, M., Pusey, A. E., and Gilby, I. C. (2014). Sexually coercive male chimpanzees sire more offspring. *Current Biology* 24: 2855–2860.

Ferguson, R. B. (2011). Born to live: Challenging killer myths. In *Origins of Altruism and Cooperation* (R. W. Sussman and C. R. Cloninger, eds.), Springer, pp. 249–270.

Ferguson, R. B. (2014). Comment on Wilson *et al.* (2014a) at http://blogs.scientificamerican.com/cross-check/2014/09/18/anthropologist-brian-ferguson-challenges-claim-that-chimp-violence-is-adaptive/

Fischer, J. and Price, T. (2016). Meaning, intention, and inference in primate communication. *Neuroscience and Biobehavioral Reviews*, https://doi.org/10.1016/j.neubiorev.2016.10.014

Fischer, J., Wheeler, B. C., and Higham, J. P. (2015). Is there any evidence for vocal learning in chimpanzee food calls? *Current Biology* 25: R1028–R1029.

Fitzpatrick, C. L., Altmann, J., and Alberts, S. C. (2015). Exaggerated sexual swellings and male mate choice in primates: Testing the reliable indicator hypothesis in the Amboseli baboons. *Animal Behaviour* 104: 175–185.

Fleagle, J. G. (2013). *Primate Adaptation and Evolution*, 3rd edn., Academic Press, San Diego, CA.

Foerster, S., McLellan, K., Schroepfer-Walker, K., Murray, C. M., Krupenye, C., Gilby, I. C., and Pusey, A. E. (2015). Social bonds in the dispersing sex: Partner preferences among adult female chimpanzees. *Animal Behaviour* 105: 139–152.

Foster, M. W., Gilby, I. C., Murray, C. M., Johnson, A., Wroblewski, E. E., and Pusey, A. E. (2009). Alpha male chimpanzee grooming patterns: Implications for dominance "style." *American Journal of Primatology* 71: 136–144.

Fourrier, M., Sussman, R. W., Kippen, R., and Childs, G. (2008). Demographic modeling of a predator-prey system and its implication for the Gombe population of *Procolobus rufomitratus tephrosceles*. *International Journal of Primatology* 29: 497–508.

Fowler, A. and Sommer, V. (2007). Subsistence technology of Nigerian chimpanzees. *International Journal of Primatology* 28: 997–1023.

Fragaszy, D. (2003). Making space for traditions. *Evolutionary Anthropology* 12: 61–70.

Fragaszy, D. M. and Perry, S., eds. (2003a). *The Biology of Traditions: Models and Evidence*. Cambridge University Press, Cambridge.

Fragaszy, D.M. and Perry, S. (2003b). Towards a biology of traditions. In *The Biology of Traditions: Models and Evidence* (D. M. Fragaszy and S. Perry, eds.), Cambridge University Press, Cambridge, pp. 1–32.

Fraser, O. N. and Aureli, F. (2008). Reconciliation, consolation and postconflict behavioral specificity in chimpanzees. *American Journal of Primatology* 70: 1114–1123.

Fraser, O. N., Stahl, D., and Aureli, F. (2008). Stress reduction through consolation in chimpanzees. *Proceedings of the National Academy of Sciences of the United States of America* 105: 8557–8562.

Fraser, O. N., Koski, S.E., Wittig, R. M., and Aureli, F. (2009). Why are bystanders friendly to recipients of aggression? *Communicative & Integrative Biology* 2: 285–291.

Fröhlich, M., Kuchenbuch, P., Müller, G., Fruth, B., Furuichi, T., Wittig, R. M., and Pika, S. (2016). Unpeeling the layers of language: Bonobos and chimpanzees engage in cooperative turn-taking sequences. *Scientific Reports* 6: 25887; doi:10.1038/srep2588

Fröhlich, M., Müller, G., Zeiträg, C., Wittig, R. M., and Pika, S. (2017). Gestural development of chimpanzees in the wild: The impact of interactional experience. *Animal Behaviour*, http://dx.doi.org/10.1016/j.anbehav.2016.12.018

Furuichi, T. (1989). Social interactions and the life history of female *Pan paniscus* in Wamba, Zaire. *International Journal of Primatology* 10(3): 173–197.

Furuichi, T. and Thompson, J., eds. (2008). *The Bonobos: Behavior, Ecology, and Conservation.* Springer, New York, NY.

Geissmann, T. (2002). Taxonomy and evolution of gibbons. *Evolutionary Anthropology* Supplement 1: 28–31.

Georgiev, A. V., Russell, A. F., Emery Thompson, M., Otali, E., Muller, M. N., and Wrangham, R. W. (2014). The foraging costs of mating effort in male chimpanzees (*Pan troglodytes schweinfurthii*). *International Journal of Primatology* 35: 725–745.

Ghiglieri, M. (1984). *The Chimpanzees of Kibale Forest: A Field Study of Ecology and Social Structure.* Columbia University Press, New York, NY.

Gilby, I. C. (2006). Meat sharing among Gombe chimpanzees: Harassment and reciprocal exchange. *Animal Behaviour* 71: 953–963.

Gilby, I. C. and Wrangham, R. W. (2007). Risk-prone hunting by chimpanzees (*Pan troglodytes schweinfurthii*) increases during periods of high diet quality. *Behavioral Ecology and Sociobiology* 61: 1771–1779.

Gilby, I. C. and Wrangham, R. W. (2008). Association patterns among wild chimpanzees (*Pan troglodytes schweinfurthii*) reflect sex differences in cooperation. *Behavioral Ecology and Sociobiology* 62: 1831–1842.

Gilby, I. C., Eberly, L. E., Pintea, L., and Pusey, A. E. (2006). Ecological and social influences on the hunting behaviour of wild chimpanzees. *Animal Behaviour* 72: 169–180.

Gilby, I. C., Eberly, L. E., and Wrangham, R. W. (2008). Economic profitability of social predation among wild chimpanzees: Individual variation promotes cooperation. *Animal Behaviour* 75: 351–360.

Gilby, I. C., Emery Thompson, M., Ruane, J. D., and Wrangham, R. W. (2010). No evidence for short-term exchange of meat for sex among chimpanzees. *Journal of Human Evolution* 59: 44–53.

Gilby, I. C., Wilson, M. L., and Pusey, A. E. (2013a). Ecology rather than psychology explains co-occurrence of predation and border patrols in male chimpanzees. *Animal Behaviour* 86: 61–74.

Gilby, I. C., Brent, L. J. N., Wroblewski, E. E., Rudicell, R. S., Hahn, B. H., Goodall, J., and Pusey, A. E. (2013b). Fitness benefits of coalitionary aggression in male chimpanzees. *Behavioral Ecology and Sociobiology* 67: 373–381.

Gilby, I. C., Machanda, Z. P., Mjungu, D. C., Rosen, J., Muller, M. N., Pusey, A. E., and Wrangham, R. W. (2015). "Impact hunters" catalyse cooperative hunting in two wild chimpanzee communities. *Philosophical Transactions of the Royal Society B* 370: 20150005.

Gilby, I. C., Machanda, Z. P., O'Malley, R. C., Murray, C. M., Lonsdorf, E. V., Walker, K., Mjungu, D. C., Otali, E., Muller, M. N., Emery Thompson, M., Pusey, A. E., and Wrangham, R. W. (2017). Predation by female

chimpanzees: toward an understanding of sex differences in meat acquisition in the last common ancestor of *Pan* and *Homo*. *Journal of Human Evolution*
110: 82–94.

Goldberg, T. L. and Wrangham, R. W. (1997). Genetic correlates of social behaviour in wild chimpanzees: Evidence from mitochondrial DNA. *Animal
Behaviour* 54: 559–570.

Gomes, C. M. and Boesch, C. (2009). Wild chimpanzees exchange meat for sex on
a long-term basis. *PLoS ONE* 4: 1–6 (e5116).

Gomes, C. M. and Boesch, C. (2011). Reciprocity and trades in wild West African
chimpanzees. *Behavioral Ecology and Sociobiology* 65: 2183–2196.

Gomes, C. M., Mundry, R., and Boesch, C. (2009). Long-term reciprocation of
grooming in wild West African chimpanzees. *Proceedings of the Royal Society B*
276: 699–706.

Goodall, J. (1965). Chimpanzees of the Gombe Stream Reserve. In *Primate
Behavior: Field Studies of Monkeys and Apes* (I. DeVore, ed.), Holt, Rinehart &
Winston, New York, NY, pp. 425–473.

Goodall, J. (1967/2005). Mother-offspring relationships in free-ranging chimpanzees. In *Primate Ethology* (D. Morris, ed.), Aldine, New Brunswick, NJ, pp.
287–346.

Goodall, J. (1968a). The behaviour of free-living chimpanzees in the Gombe
Stream Reserve. *Animal Behaviour Monographs* 1: 163–311.

Goodall, J. (1968b). A preliminary report on expressive movements and communication in the Gombe Stream chimpanzees. In *Primate Patterns* (P.C. Jay,
ed.), Holt, Rinehart & Winston, New York, NY, pp. 313–374.

Goodall, J. (1970). Tool-using in primates and other vertebrates. In *Advances in the
Study of Behavior, Vol. 3* (R. A. Hinde and E. Shaw, eds.), Academic Press,
New York, NY, pp. 195–249.

Goodall, J. (1971). *In the Shadow of Man*, 3rd edn. Houghton Mifflin Harcourt,
New York, NY.

Goodall, J. (1973). Cultural elements in a chimpanzee community. In *Precultural
Primate Behaviour, Symposia of the 4th International Congress of Primatology, Vol. 1*
(E. W. Menzel, ed.), Karger, Basel, pp. 144–184.

Goodall, J. (1977). Infant-killing and cannibalism in free-living chimpanzees.
Folia Primatologica 28: 259–282.

Goodall, J. (1986). *The Chimpanzees of Gombe*. Belknap Press of Harvard University
Press, Cambridge, MA.

Goodall, J. (1992). Unusual violence in the overthrow of an alpha male chimpan-
zee at Gombe. In *Topics in Primatology, Vol. 1: Human Origins* (T. Nishida,
W. C. McGrew, P. Marler, M. Pickford and F. B. M. de Waal, eds.),
University of Tokyo Press, Tokyo, pp. 131–142.

Goodall, J., Bandora, A., Bergmann, E., Busse, C., Matama, H., Mpongo, E.,
Pierce, A., and Riss, D. (1979). Intercommunity interactions in the chimpanzee population of the Gombe National Park. In *The Great Apes* (D. A. Hamburg
and E. R. McCown, eds.), Benjamin/Cummings, Menlo Park, CA, pp. 13–53.

Goosen, C. (1987). Social grooming in primates. In *Comparative Primate Biology
Vol. 2, Part B: Behavior, Cognition and Motivation* (G. Mitchell and J. Erwin, eds.),
Alan R. Liss, New York, NY, pp. 107–132.

Gradstein, F. M. (2012). *A Geologic Time Scale*. Oxford University Press, Oxford.

Graham, C. E. (1981). Menstrual cycle of the great apes. In *Reproductive Biology of
the Great Apes* (C. E. Graham, ed.), Academic Press, New York, pp. 1–43.

Groves, C. (2001). *Primate Taxonomy*. Smithsonian Institution Press, Washington,
D.C.

Gruber, T. and Zuberbühler, K. (2013). Vocal recruitment for joint travel in wild chimpanzees. *PLoS ONE* 8(9): e76073. doi:10.1371/journal.pone.0076073

Grueter, C. C. (2015). Home range overlap as a driver of intelligence in primates. *American Journal of Primatology* 77: 418–424.

Halperin, S. D. (1979). Temporary association patterns in free ranging chimpanzees. In *The Great Apes* (D. A. Hamburg and E. R. McCown, eds.), Benjamin/Cummings, Menlo Park, CA, pp. 491–499.

Hamai, M., Nishida, T., Takasaki, H., and Turner, L. A. (1992). New records of within-group infanticide in wild chimpanzees. *Primates* 33(2): 151–162.

Hamilton, W. D. (1964). The genetical evolution of social behavior I and II. *Journal of Theoretical Biology* 7: 1–52.

Harcourt, A. H. and Stewart, K. J. (2007). *Gorilla Society*. University of Chicago Press, Chicago, IL.

Hartwig, W. C. (2002). *The Primate Fossil Record*. Cambridge University Press, Cambridge.

Hasegawa, T. (1989). Sexual behavior of immigrant and resident female chimpanzees at Mahale. In *Understanding Chimpanzees* (P. G. Heltne and L. A. Marquardt, eds.), Harvard University Press, Cambridge, MA, pp. 90–103.

Hasegawa, T. (1990). Sex differences in ranging patterns. In *The Chimpanzees of the Mahale Mountains* (T. Nishida, ed.), University of Tokyo Press, Tokyo, pp. 99–114.

Hasegawa, T. and Hiraiwa-Hasegawa, M. (1983). Opportunistic and restrictive matings among wild chimpanzees in the Mahale mountains, Tanzania. *Journal of Ethology* 1: 75–85.

Hasegawa, T. and Hiraiwa-Hasegawa, M. (1990). Sperm competition and mating behavior. In *The Chimpanzees of the Mahale Mountains*. (T. Nishida, ed.), University of Tokyo Press, Tokyo, pp. 115–132.

Hashimoto, C. and Furuichi, T. (2006). Frequent copulation by females and high promiscuity in chimpanzees in the Kalinzu Forest, Uganda. In *Primates of Western Uganda* (N. Newton-Fisher, H. Notman, V. Reynolds, and J.D. Paterson, eds.), Springer, New York, NY, pp. 247–257.

Hayaki, H. (1985). Social play of juvenile and adolescent chimpanzees in the Mahale Mountains National Park, Tanzania. *Primates* 26: 343–360.

Hayaki, H. (1988). Association partners of young chimpanzees in the Mahale Mountains National Park, Tanzania. *Primates* 29: 147–161.

Hayaki, H. (1990). Social context of pant-grunting in young chimpanzees. In *The Chimpanzees of the Mahale Mountains* (T. Nishida, ed.), University of Tokyo Press, Tokyo, pp. 189–206.

Hayaki, H., Huffman, M. A., and Nishida, T. (1989). Dominance among male chimpanzees in the Mahale Mountains National Park, Tanzania: A preliminary study. *Primates* 30: 187–197.

Henzi, S. P. and Barrett, L. (1999). The value of grooming to female primates. *Primates* 40: 47–59.

Herbinger, I., Papworth, S., Boesch, C., and Zuberbühler, K. (2009). Vocal, gestural and locomotor responses of wild chimpanzees to familiar and unfamiliar intruders: A playback study. *Animal Behaviour* 78: 1389–1396.

Hey, J. (2010). The divergence of chimpanzee species and subspecies as revealed in multipopulation isolation-with-migration analysis. *Molecular Biology and Evolution* 27: 921–933.

Hildebrand, M. (1977). Analysis of asymmetrical gaits. *Journal of Mammalogy* 58: 131–156.

Hill, K. (2009). Animal "culture"? In *The Question of Animal Culture* (K. N. Laland and B. G. Galef, eds.), Harvard University Press, Cambridge, MA, pp. 269–287.

Hinde, R. A. (1976). Interactions, relationships and social structure. *Man* 11: 1–17.

Hiraiwa-Hasegawa, M. (1989). Sex differences in the behavioral development of chimpanzees at Mahale. In *Understanding Chimpanzees* (P. G. Heltne and L. A. Marquardt, eds.), Harvard University Press, Cambridge, MA, pp. 104–115.

Hiraiwa-Hasegawa, M. (1992). Cannibalism among non-human primates. In *Cannibalism: Ecology and Evolution among Diverse Taxa* (M. A. Elgar and B. J. Crespi, eds.), Oxford University Press, Oxford, pp. 323–338.

Hiraiwa-Hasegawa, M., Hasegawa, T., and Nishida, T. (1984). Demographic study of a large-sized unit-group of chimpanzees in the Mahale Mountains, Tanzania: A preliminary report. *Primates* 25: 401–413.

Hirata, S., Yamakoshi, G., Fujita, S., Ohashi, G., and Matsuzawa, T. (2001). Capturing and toying with hyraxes (*Dendrohyrax dorsalis*) by wild chimpanzees (*Pan troglodytes*) at Bossou, Guinea. *American Journal of Primatology* 53: 93–97.

Hobaiter, C. and Byrne, R. W. (2011a). The gestural repertoire of the wild chimpanzee. *Animal Cognition* 14: 745–767.

Hobaiter, C. and Byrne, R. W. (2011b). Serial gesturing by wild chimpanzees: Its nature and function for communication. *Animal Cognition* 14: 827–838.

Hobaiter, C., Schel, A. M., Langergraber, K., and Zuberbühler, K. (2014a). "Adoption" by maternal siblings in wild chimpanzees. *PLoS ONE* 9(8): 1–6 (e103777).

Hobaiter, C., Poisot, T., Zuberbühler, K., Hoppitt, W., and Gruber, T. (2014b). Social network analysis shows direct evidence for social transmission of tool use in wild chimpanzees. *PLoS Biology* 12: 1–12.

Hockings, K. J. (2011). The crop-raiders of the sacred hill. In *The Chimpanzees of Bossou and Nimba* (T. Matsuzawa, T. Humle, and Y. Sugiyama, eds.), Springer, Tokyo, pp. 211–220.

Hockings, K. J., Yamakoshi, G., Kabasawa, A., and Matsuzawa, T. (2010). Attacks on local persons by chimpanzees in Bossou, Republic of Guinea: Long-term perspectives. *American Journal of Primatology* 72: 887–896.

Hopkins, W. D., Taglialatela, J. P., and Leavens, D. A. (2007). Chimpanzees differentially produce novel vocalizations to capture the attention of a human. *Animal Behaviour* 73: 281–286.

Hosaka, K. (2015a). Hunting and food sharing. In *Mahale Chimpanzees: 50 Years of Research* (M. Nakamura, K. Hosaka, N. Itoh, and K. Zamma, eds.), Cambridge University Press, Cambridge, pp. 274–290.

Hosaka, K. (2015b). Intimidation display. In *Mahale Chimpanzees: 50 Years of Research* (M. Nakamura, K. Hosaka, N. Itoh, and K. Zamma, eds.), Cambridge University Press, Cambridge, pp. 435–447.

Hosaka, K., Nishida, T., Hamai, M., Matsumoto-Oda, A., and Uehara, S. (2001). Predation of mammals by the chimpanzees of the Mahale Mountains, Tanzania. In *All Apes Great and Small Vol. 1: African Apes* (B. M. F. Galdikas, N. E. Briggs, L. K. Sheeran, G. L. Shapiro, and J. Goodall, eds.), Kluwer Academic, NY, pp. 107–130.

Hrdy, S. B. (1974). Male–male competition and infanticide among the langurs (*Presbytis entellus*) of Abu, Rajasthan. *Folia Primatologica* 22: 19–58.

Hrdy, S. B. (1977). *The Langurs of Abu*. Harvard University Press, Cambridge, MA.

Hrdy, S. B. (1979). Infanticide among animals: A review, classification, and examination of the implications for the reproductive strategies of females. *Ethology and Sociobiology* 1: 13–40.

Hrdy, S. B. (1981). *The Woman That Never Evolved*. Harvard University Press, Cambridge, MA.

Huchard, E., Courtiol, A., Benavides, J. A., Knapp, L. A., Raymond, M., and Cowlishaw, G. (2009). Can fertility signals lead to quality signals? Insights from the evolution of primate sexual swellings. *Proceedings of the Royal Society B* 276: 1889–1897.

Huffman, M. A. (1990). Some socio-behavioral manifestations of old age. In *The Chimpanzees of the Mahale Mountains* (T. Nishida, ed.). University of Tokyo Press, Tokyo, pp. 236–255.

Huffman, M. A. (2015). Chimpanzee self-medication: A historical perspective of the key findings. In *Mahale Chimpanzees: 50 Years of Research* (M. Nakamura, K. Hosaka, N. Itoh, and K. Zamma, eds.), Cambridge University Press, Cambridge, pp. 340–353.

Huffman, M. A. and Hirata, S. (2004). An experimental study of leaf swallowing in captive chimpanzees: Insights into the origin of a self-medicative behavior and the role of social learning. *Primates* 45: 113–118.

Huffman, M. A. and Kalunde, M. A. (1993). Tool-assisted predation on a squirrel by a female chimpanzee in the Mahale Mountains, Tanzania. *Primates* 34: 93–98.

Huffman, M. A., Spiezio, C., Sgaravatti, A., and Leca, J. (2010). Leaf swallowing behavior in chimpanzees (*Pan troglodytes*): Biased learning and the emergence of group level cultural differences. *Animal Cognition* 13: 871–880.

Humle, T. (2011). Location and ecology. In *The Chimpanzees of Bossou and Nimba* (T. Matsuzawa, T. Humle, and Y. Sugiyama, eds.), Springer, Tokyo, pp. 13–21.

Humle, T. and Fragaszy, D. M. (2011). Tool use and cognition in primates. In *Primates in Perspective* (C. J. Campbell, A. Fuentes, K. C. MacKinnon, S. K. Bearder, and R. M. Stumpf, eds.), Oxford University Press, Oxford, pp. 637–651.

Humle, T., Snowdon, C. T., and Matsuzawa, T. (2009a). Social influences on ant-dipping acquisition in the wild chimpanzees (*Pan troglodytes verus*) of Bossou, Guinea, West Africa. *Animal Cognition* 12 (Suppl. 1): S37–S48.

Humle, T., Colin, C., and Raballand, E. (2009b). Preliminary report on hand-clasp grooming in sanctuary-released chimpanzees, Haut Niger National Park, Guinea. *Pan African News* 16: 1–3.

Humphrey, N. K. (1976). The social function of intellect. In *Growing Points in Ethology* (R. Hinde and P. Bateson, eds.), Cambridge University Press, Cambridge, pp. 303–317.

Hunt, K. D. (1989). Positional behavior in *Pan troglodytes* at the Mahale Mountains and the Gombe Stream National Parks, Tanzania. Ph.D. dissertation, University of Michigan, Ann Arbor.

Hyeroba, D., Apell, P., and Otali, E. (2011). Managing a speared alpha male chimpanzee (*Pan troglodytes*) in Kibale National Park, Uganda. *Veterinary Record* 169: 658. doi:10.1136/vr.d4680

Inoue-Nakamura, N. and Matsuzawa, T. (1997). Development of stone tool use by wild chimpanzees (*Pan troglodytes*). *Journal of Comparative Psychology* 111: 159–173.

Inoue, E., Inoue-Murayama, M., Vigilant, L., Takenaka, O., and Nishida, T. (2008). Relatedness in wild chimpanzees: Influence of paternity, male philopatry, and demographic factors. *American Journal of Physical Anthropology* 137: 256–262.

Isabirye-Basuta, G. (1989). The Ecology and Conservation Status of the Chimpanzee *Pan troglodytes schweinfurthii* in Kibale Forest, Uganda. Ph. D. dissertation, Makerere University.

Itoh, N. and Muramatsu, D. (2015). Patterns and trends in fruiting phenology: Some important implications for chimpanzee diet. In *Mahale Chimpanzees: 50*

Years of Research (M. Nakamura, K. Hosaka, N. Itoh, and K. Zamma, eds.), Cambridge University Press, Cambridge, pp. 174–194.

Itoh, N. and Nakamura, M. (2015). Diet and feeding behavior. In *Mahale Chimpanzees: 50 Years of Research* (M. Nakamura, K. Hosaka, N. Itoh, and K. Zamma, eds.), Cambridge University Press, Cambridge, pp. 227–245.

IUCN/SSC Primate Specialist Group (2017). www.primate-sg.org/species.

Janmaat, K. R. L., Ban, S. D., and Boesch, C. (2013). Chimpanzees use long-term spatial memory to monitor large fruit trees and remember feeding experiences across seasons. *Animal Behaviour* 86: 1183–1205.

Janmaat, K. R. L., Polansky, L., Ban, S. D., and Boesch, C. (2014). Wild chimpanzees plan their breakfast time, type, and location. *Proceedings of the National Academy of Sciences U.S.A.* 111: 16343–16348.

Jenny, D. and Zuberbühler, K. (2005). Hunting behaviour of West African forest leopards. *African Journal of Ecology* 43: 197–200.

Jensen, K. (2012). Social regard: Evolving a psychology of cooperation. In *The Evolution of Primate Societies* (J. C. Mitani, J. Call, P. M. Kappeler, R. A. Palombit, and J. B. Silk, eds.), University of Chicago Press, Chicago, IL, pp. 565–584.

Jones, J. H., Wilson, M. L., Murray, C., and Pusey, A. (2010). Phenotypic quality influences fertility in Gombe chimpanzees. *Journal of Animal Ecology* 79: 1262–1269.

Judge, P. G. (1991). Dyadic and triadic reconciliation in pigtail macaques (*Macaca nemestrina*). *American Journal of Primatology* 23: 225–237.

Kaas, J. H. (2005). The evolution of visual cortex in primates. In *The Primate Visual System* (J. Kremers, ed.), John Wiley & Sons, New York, NY, pp. 267–283.

Kaburu, S. S. K. and Newton-Fisher, N. E. (2013). Social instability raises the stakes during social grooming among wild male chimpanzees. *Animal Behaviour* 86: 519–527.

Kaburu, S. S. K. and Newton-Fisher, N. E. (2015a). Egalitarian despots: Hierarchy steepness, reciprocity and the grooming-trade model in wild chimpanzees, *Pan troglodytes. Animal Behaviour* 99: 61–71.

Kaburu, S. S. K. and Newton-Fisher, N. E. (2015b). Trading or coercion? Variation in male mating strategies between two communities of East African chimpanzees. *Behavioral Ecology and Sociobiology* 69: 1039–1052.

Kaburu, S. S. K., Inoue, S., and Newton-Fisher, N. E. (2013). Death of the alpha: Within-community lethal violence among chimpanzees of the Mahale Mountains National Park. *American Journal of Primatology* 75: 789–797.

Kahlenberg, S. M. and Wrangham, R. W. (2010). Sex differences in chimpanzees' use of sticks as play objects resemble those of children. *Current Biology* 20: R1067–1068.

Kahlenberg, S. M., Emery Thompson, M., and Wrangham, R. W. (2008a). Female competition over core areas in *Pan troglodytes schweinfurthii*, Kibale National Park, Uganda. *International Journal of Primatology* 29: 931–947.

Kahlenberg, S. M., Emery Thompson, M., Muller, M. N., and Wrangham, R. W. (2008b). Immigration costs for female chimpanzees and male protection as an immigrant counterstrategy to intrasexual aggression. *Animal Behaviour* 76: 1497–1509.

Kalan, A. K., Mundry, R., and Boesch, C. (2015). Wild chimpanzees modify food call structure with respect to tree size for a particular fruit species. *Animal Behaviour* 101: 1–9.

Kamilar, J. M. and Marshack, J. L. (2012). Does geography or ecology best explain "cultural" variation among chimpanzee communities? *Journal of Human Evolution* 62: 256–260.

Kappeler, P. M. (2012). The behavioral ecology of strepsirrhines and tarsiers. In *The Evolution of Primate Societies* (J. C. Mitani, J. Call, P. M. Kappeler, R. A. Palombit, and J. B. Silk, eds.), University of Chicago Press, Chicago, IL, pp. 19–42.

Kappeler, P. M. and van Schaik, C. P. (2002). Evolution of primate social systems. *International Journal of Primatology* 23: 707–740.

Kawanaka, K. (1989). Age differences in social interactions of young males in a chimpanzee unit-group at the Mahale Mountains National Park, Tanzania. *Primates* 30(3): 285–305.

Kawanaka, K. (1990). Alpha males' interactions and social skills. In *The Chimpanzees of the Mahale Mountains* (T. Nishida, ed.), University of Tokyo Press, Tokyo, pp. 171–187.

Kawanaka, K. and Nishida, T. (1974). Recent advances in the study of inter-unit-group relationships and social structure of wild chimpanzees of the Mahali Mountains. In *Proceedings of the Symposia of the 5th Congress of the International Primatological Society* (S. Kondo, A. Ehara, M. Kawai, and S. Kawamura, eds.), Japan Science Press, Tokyo, pp. 173–186.

Knott, C. (2001). Female reproductive ecology of the apes: Implications for human evolution. In *Reproductive Ecology and Human Evolution* (P.T. Ellison, ed.), Aldine de Gruyter, New York, NY, pp. 429–463.

Kojima, S. (2001). Early vocal development in a chimpanzee. In *Primate Origins of Human Cognition and Behavior* (T. Matsuzawa, ed.), Springer, Hong Kong, pp. 190–196.

Kojima, S., Izumi, A., and Ceugniet, M. (2003). Identification of vocalizers by pant hoots, pant grunts and screams in a chimpanzee. *Primates* 44: 225–230.

Köndgen, S., Kühl, H., N'Goran, P. K., Walsh, P. D., Schenk, S., Ernst, N., Biek, R., Formenty, P., Matz-Rensing, K., Schweiger, B., Junglen, S., Ellerbrok, H., Nitsche, A., Briese, T., Lipkin, W. I., Pauli, G., Boesch, C., and Leendertz, F. H. (2008). Pandemic human viruses cause decline of endangered great apes. *Current Biology* 18: 1–5.

Koops, K., Humle, T., Sterck, E. H. M., and Matsuzawa, T. (2007). Ground nesting by the chimpanzees of the Nimba Mountains, Guinea: Environmentally or socially determined? *American Journal of Primatology* 69: 407–419.

Koops, K., McGrew, W. C., de Vries, H., and Matsuzawa, T. (2012). Nest-building by chimpanzees (*Pan troglodytes verus*) at Seringbara, Nimba Mountains: Antipredation, thermoregulation, and antivector hypotheses. *International Journal of Primatology* 33: 356–380.

Koops, K., Schöning. C., Isaji, M., and Hashimoto, C. (2015). Cultural differences in ant-dipping tool length between neighbouring chimpanzee communities at Kalinzu, Uganda. *Scientific Reports* 5: 12456; doi:10.1038/srep12456

Koski, S. E. and Sterck, E. H. M. (2007). Triadic postconflict affiliation in captive chimpanzees: Does consolation console? *Animal Behaviour* 73: 133–142.

Koski, S. E. and Sterck, E. H. M. (2009). Post-conflict third-party affiliation in chimpanzees: What's in it for the third party? *American Journal of Primatology* 71: 409–418.

Kraft, T., Venkataraman, V. V., and Dominy, N. J. (2014). A natural history of human tree climbing. *Journal of Human Evolution* 71: 105–118.

Kroeber, A. L. (1928). Sub-human culture beginnings. *Quarterly Review of Biology* 3: 325–342.

Kühl, H., Sop, T., Williamson, E. A., Mundry, R., Brugière, D., Campbell, G., Cohen, H., Danquah, E., Ginn, L., Herbinger, I., Jones, S., Junker, J., Kormos, R., Kouakou, C. Y., N'Goran, P. K., Normand, E., Shutt-Phillips, K., Tickle, A., Vendras, E., Welsh, A., Wessling, E. G., and Boesch, C. (2017).

The critically endangered western chimpanzee declines by 80%. *American Journal of Primatology.* https://doi.org/10.1002/ajp.22681

Kummer, H. (1968). Tripartite relations in hamadryas baboons. In *Social Communication among Primates* (S. A. Altmann, ed.), University of Chicago Press, Chicago, IL, pp. 63–72.

Kummer, H. (1978). On the value of social relationships to nonhuman primates: A heuristic scheme. *Social Science Information* 17: 687–705.

Kummer, H. (1995). *In Quest of the Sacred Baboon.* Princeton University Press, Princeton, NJ.

Kutsukake, N. and Castles, D. L. (2004). Reconciliation and post-conflict third-party affiliation among wild chimpanzees in the Mahale Mountains, Tanzania. *Primates* 45: 157–165.

Kutsukake, N. and Matsusaka, T. (2002). Incident of intense aggression by chimpanzees against an infant from another group in Mahale Mountains National Park, Tanzania. *American Journal of Primatology* 58: 175–180.

Laland, K. N. and Galef, B. G., eds. (2009). *The Question of Animal Culture.* Harvard University Press, Cambridge, MA.

Lamon, N., Neumann, C., Gruber, T., and Zuberbühler, K. (2017). Kin-based cultural transmission of tool use in wild chimpanzees. *Science Advances* 3: e1602750.

Langergraber, K. E. and Vigilant, L. (2011). Genetic differences cannot be excluded from generating behavioural differences among chimpanzee groups. *Proceedings of the Royal Society B* 278: 2094–2095. doi:10.1098/rspb.2011.0391

Langergraber, K., Mitani, J., and Vigilant, L. (2007). The limited impact of kinship on cooperation in wild chimpanzees. *Proceedings of the National Academy of Sciences U.S.A.* 104(19): 7786–7790.

Langergraber, K., Mitani, J., and Vigilant, L. (2009). Kinship and social bonds in female chimpanzees (*Pan troglodytes*). *American Journal of Primatology* 71: 840–851.

Langergraber, K. E., Boesch, C., Inoue, E., Inoue-Murayama, M., Mitani, J. C., Nishida, T., Pusey, A., Reynolds, V., Schubert, G., Wrangham, R. W., Wroblewski, E., and Vigilant, L. (2011). Genetic and "cultural" similarity in wild chimpanzees. *Proceedings of the Royal Society B* 278: 408–416. doi:10.1098/rspb.2010.1112

Langergraber, K., Mitani, J., Watts, D. P., and Vigilant, L. (2013). Male–female socio-spatial relationships and reproduction in wild chimpanzees. *Behavioral Ecology and Sociobiology* 67: 861–873.

Langergraber, K. E., Rowney, C., Crockford, C., Wittig, R., Zuberbühler, K., and Vigilant, L. (2014). Genetic analyses suggest no immigration of adult females and their offspring into the Sonso community of chimpanzees in the Budongo Forest Reserve, Uganda. *American Journal of Primatology* 76: 640–648.

LaPorte, M. N. C. and Zuberbühler, K. (2010). Vocal greeting behaviour in wild chimpanzee females. *Animal Behaviour* 80: 467–473.

Lee, P. C. (1987). Nutrition, fertility and maternal investment in primates. *Journal of Zoology* 213: 409–422.

Lehmann, J. and Boesch, C. (2004). To fission or to fusion: Effects of community size on wild chimpanzee (*Pan troglodytes verus*) social organization. *Behavioral Ecology and Sociobiology* 56: 207–216.

Lehmann, J. and Boesch, C. (2005). Bisexually bonded ranging in chimpanzees (*Pan troglodytes verus*). *Behavioral Ecology and Sociobiology* 57: 525–535.

Lehmann, J. and Boesch, C. (2008). Sexual differences in chimpanzee sociality. *International Journal of Primatology* 29: 65–81.

Lehmann, J. and Boesch, C. (2009). Sociality of the dispersing sex: The nature of social bonds in West African female chimpanzees, *Pan troglodytes*. *Animal Behaviour* 77: 377–387.

Lehmann, J., Fickenscher, G., and Boesch, C. (2006). Kin biased investment in wild chimpanzees. *Behaviour* 143: 931–955.

Lehmann, J., Korstjens, A. H., and Dunbar, R. I. M. (2007). Group size, grooming and social cohesion in primates. *Animal Behaviour* 74: 1617–1629.

Levréro, F. and Mathevon, N. (2013). Vocal signature in wild infant chimpanzees. *American Journal of Primatology* 75: 324–332.

Lonsdorf, E. V. (2005). Sex differences in the development of termite-fishing skills in the wild chimpanzees, *Pan troglodytes schweinfurthii*, of Gombe National Park, Tanzania. *Animal Behaviour* 70: 673–683.

Lonsdorf, E. V. (2006). What is the role of mothers in the acquisition of termite-fishing behaviors in wild chimpanzees (*Pan troglodytes schweinfurthii*)? *Animal Cognition* 9: 36–46.

Lonsdorf, E. V., Eberly, L. E., and Pusey, A. E. (2004). Sex differences in learning in chimpanzees. *Nature* 428: 715–716.

Lonsdorf, E. V., Ross, S. R., and Matsuzawa, T., eds. (2010). *The Mind of the Chimpanzee*. University of Chicago Press, Chicago, IL.

Lonsdorf, E. V., Anderson, K. E., Stanton, M. A., Shender, M., Heintz, M. R., Goodall, J., and Murray, C. M. (2014a). Boys will be boys: Sex differences in wild chimpanzee social interactions. *Animal Behaviour* 88: 79–83.

Lonsdorf, E. V., Markham, A. C., Heintz, M. R., Anderson, K. E., Ciuk, D. J., Goodall, J., and Murray, C. M. (2014b). Sex differences in wild chimpanzee behavior emerge during infancy. *PLoS ONE* 9: e99099. doi:10.1371/journal.pone.0099099

Luncz, L. V. and Boesch, C. (2014). Tradition over trend: Neighboring chimpanzee communities maintain differences in cultural behavior despite frequent immigration of adult females. *American Journal of Primatology* 76: 649–657.

Luncz, L. V. and Boesch, C. (2015). The extent of cultural variation between adjacent chimpanzee (*Pan troglodytes verus*) communities: A microecological approach. *American Journal of Physical Anthropology* 156: 67–75.

Luncz, L. V., Mundry, R., and Boesch, C. (2012). Evidence for cultural differences between neighboring chimpanzee communities. *Current Biology* 22: 922–926.

Lwanga, J. S., Struhsaker, T. T., Struhsaker, P. J., Butynski, T. M., and Mitani, J. C. (2011). Primate population dynamics over 32.9 years at Ngogo, Kibale National Park, Uganda. *American Journal of Primatology* 73: 997–1011.

Lycett, S. J., Collard, M., and McGrew, W. C. (2007). Phylogenetic analyses of behavior support existence of culture among wild chimpanzees. *Proceedings of the National Academy of Sciences U.S.A.* 104: 17588–17592.

Lycett, S. J., Collard, M., and McGrew, W. C. (2009). Cladistic analyses of behavioural variation in wild *Pan troglodytes*: Exploring the chimpanzee culture hypothesis. *Journal of Human Evolution* 142: 337–349.

Lycett, S. J., Collard, M., and McGrew, W. C. (2010). Are behavioral differences among wild chimpanzee communities genetic or cultural? An assessment using tool-use data and phylogenetic methods. *American Journal of Physical Anthropology* 142: 461–467.

Macedonia, J. M. and Evans, C. S. (1993). Variation among mammalian alarm call systems and the problem of meaning in animal signals. *Ethology* 93: 177–197.

Machanda, Z. P., Gilby, I. C., and Wrangham, R. W. (2013). Male–female association patterns among free-ranging chimpanzees (*Pan troglodytes schweinfurthii*). *International Journal of Primatology* 34: 917–938.

Machanda, Z. P., Gilby, I. C., and Wrangham, R. W. (2014). Mutual grooming among adult male chimpanzees: The immediate investment hypothesis. *Animal Behaviour* 87: 165–174.

Manson J. and Wrangham, R. W. (1991). Intergroup aggression in chimpanzees and humans. *Current Anthropology* 32(4): 369–390.

Marchesi, P., Marchesi, N., Fruth, B., and Boesch, C. (1995). Census and distribution of chimpanzees in Côte d'Ivoire. *Primates* 36(4): 591–607.

Markham, A. C., Santymire, R. M., Lonsdorf, E. V., Heintz, M. R., Lipende, I., and Murray, C. M. (2014). Rank effects on social stress in lactating chimpanzees. *Animal Behaviour* 87: 195–202.

Marks, J. (2002). *What It Means to Be 98% Chimpanzee*. University of California Press, Berkeley, CA.

Marler, P. (1969). Vocalizations of wild chimpanzees: An introduction. *Proceedings of the 2nd International Congress of Primatology, Atlanta, 1968, Vol. 1*, Karger, Basel, pp. 94–100.

Marler, P. (1976). Social organization, communication, and graded signals: The chimpanzee and the gorilla. In *Growing Points in Ethology* (R. Hinde and P. Bateson, eds.), Cambridge University Press, Cambridge, pp. 239–280.

Marler, P. and Hobbett, L. (1975). Individuality in a long-range vocalization of wild chimpanzees. *Zeitschrift für Tierpsychologie* 38: 97–109.

Marler, P. and Tenaza, R. (1977). Signalling behavior of apes with special reference to vocalization. In *How Animals Communicate* (T. A. Sebeok, ed.), Indiana University Press, Bloomington, pp. 965–1032.

Marler, P., Evans, C. S., and Hauser, M. D. (1992). Animal signals: Motivational, referential, or both? In *Nonverbal Vocal Communication* (H. Papousek, U. Jürgens, and M. Papousek, eds.), Cambridge University Press, Cambridge, pp. 66–86.

Marshall, A., Wrangham, R. W., and Arcadi, A. Clark (1999). Does learning affect the structure of vocalizations in chimpanzees? *Animal Behaviour* 58: 825–830.

Martin, P. and Bateson, P. (1995). *Measuring Behaviour: An Introductory Guide*. Cambridge University Press, Cambridge.

Matsumoto-Oda, A. (1999a). Mahale chimpanzees: Grouping patterns and cycling females. *American Journal of Primatology* 47: 197–207.

Matsumoto-Oda, A. (1999b). Female choice in the opportunistic mating of wild chimpanzees (*Pan troglodytes schweinfurthii*) at Mahale. *Behavioral Ecology and Sociobiology* 46: 258–266.

Matsumoto-Oda, A. and Ihara, Y. (2011). Estrous asynchrony causes low birth rates in wild chimpanzees. *American Journal of Primatology* 73: 180–188.

Matsumoto-Oda, A. and Oda, R. (1998). Changes in the activity budget of cycling female chimpanzees. *American Journal of Primatology* 46: 157–166.

Matsumoto-Oda, A., Hosaka, K., Huffman, M. A., and Kawanaka, K. (1998). Factors affecting party size in chimpanzees of the Mahale Mountains. *International Journal of Primatology* 19(6): 999–1011.

Matsumoto-Oda, A., Kutsukaka, N., Hosaka, K., and Matsusaka, T. (2007). Sniffing behaviors of Mahale chimpanzees. *Primates* 48: 81–85.

Matsusaka, T. (2004). When does play panting occur during social play in wild chimpanzees? *Primates* 45: 221–229.

Matsusaka, T., Shimada, M., and Nakamura, M. (2015). Diversity of play. In *Mahale Chimpanzees: 50 Years of Research* (M. Nakamura, K. Hosaka,

N. Itoh, and K. Zamma, eds.), Cambridge University Press, Cambridge, pp. 544–555.

Matsuzawa, T. (1994). Field experiments on use of stone tools by chimpanzees in the wild. In *Chimpanzee Cultures* (R. W. Wrangham, W. C. McGrew, F. B. M. de Waal, and P. G. Heltne, eds.), Harvard University Press, Cambridge, MA, pp. 351–370.

Matsuzawa, T. (1996). Chimpanzee intelligence in nature and captivity: Isomorphism of symbol use and tool use. In *Great Ape Societies* (W. C. McGrew, L. F. Marchant, and T. Nishida, eds.), Cambridge University Press, Cambridge, pp. 196–209.

Matsuzawa, T. (2011). Log doll: Pretence in wild chimpanzees. In *The Chimpanzees of Bossou and Nimba* (T. Matsuzawa, T. Humle, and Y. Sugiyama, eds.), Springer, Tokyo, pp. 131–135.

Matsuzawa, T., Humle, T., and Sugiyama, Y., eds. (2011). *The Chimpanzees of Bossou and Nimba*. Springer, Tokyo.

McGrew, W. C. (1977). Socialization and object manipulation of wild chimpanzees. In *Primate Bio-Social Development: Biological, Social and Ecological Determinants* (F. E. Poirer and S. Chevalier-Skolnikoff, eds.), Garland Press, New York, NY, pp. 261–288.

McGrew, W. C. (1979). Evolutionary implications of sex differences in chimpanzee predation and tool use. In *The Great Apes* (D. A. Hamburg and E. R. McCown, eds.), Benjamin/Cummings, Menlo Park, CA, pp. 440–463.

McGrew, W. C. (1992). *Chimpanzee Material Culture*. Cambridge University Press, Cambridge.

McGrew, W. C. (2004). *The Cultured Chimpanzee: Reflections on Cultural Primatology*. Cambridge University Press, Cambridge.

McGrew, W. C. (2017). Field studies of *Pan troglodytes* reviewed and comprehensively mapped, focussing on Japan's contribution to cultural primatology. *Primates* 58: 237–258.

McGrew, W. C. and Tutin, C. E. G. (1978). *Evidence for a social custom in wild chimpanzees? Man (New Series)* 13: 234–251.

McGrew, W. C., Baldwin, P. J., and Tutin, C. E. G. (1988). Diet of Wild Chimpanzees (*Pan troglodytes verus*) at Mt. Assirik, Senegal: I. Composition. *American Journal of Primatology* 16: 213–226.

McGrew, W. C., Marchant, L. F., and Nishida, T., eds. (1996). *Great Ape Societies*. Cambridge University Press, Cambridge.

McGrew, W. C., Marchant, L. F., Scott, S. E., and Tutin, C. E. G. (2001). Intergroup differences in a social custom of wild chimpanzees: The grooming hand-clasp of the Mahale Mountains. *Current Anthropology* 42: 148–153.

McGrew, W. C., Marchant, L. F., and Hunt, K. D. (2007). Etho-archaeology of manual laterality: Well digging by wild chimpanzees. *Folia Primatologica* 78: 240–244.

McLennan, M. R. (2011). Tool-use to obtain honey by chimpanzees at Bulindi: New record from Uganda. *Primates* 52: 315–322.

Mech, L. D., Adams, L. G., Meier, T. J., Burch, J. W., and Dale, B. W. (1998). *The Wolves of Denali*. University of Minnesota Press, Minneapolis, MN.

Miller, J. A., Pusey, A. E., Gilby, I. C., Schroepfer-Walker, K. Markham, A. C., and Murray, C. M. (2014). Competing for space: Female chimpanzees are more aggressive inside than outside their core areas. *Animal Behaviour* 87: 147–152.

Milton, K. (1993). Diet and primate evolution. *Scientific American* August: 86–93.

Mitani, J. C. (2006a). Demographic influences on the behavior of chimpanzees. *Primates* 47: 6–13.

Mitani, J. C. (2006b). Reciprocal exchange in chimpanzees and other primates. In *Cooperation in Primates and Humans* (P. M. Kappeler and C. P. van Schaik, eds.), Springer, Berlin, pp. 107–119.

Mitani, J. C. (2009a). Male chimpanzees form enduring and equitable social bonds. *Animal Behaviour* 77: 633–640.

Mitani, J. C. (2009b). Cooperation and competition in chimpanzees: Current understanding and future challenges. *Evolutionary Anthropology* 18: 215–227.

Mitani, J. C. and Amsler, S. J. (2003). Social and spatial aspects of male subgrouping in a community of wild chimpanzees. *Behaviour* 140: 869–884.

Mitani, J. C. and Brandt, K. L. (1994). Social factors influence the acoustic variability in the long-distance calls of male chimpanzees. *Ethology* 96: 233–252.

Mitani, J. C. and Gros-Louis, J. (1995). Species and sex differences in the screams of chimpanzees and bonobos. *International Journal of Primatology* 16: 393–411.

Mitani, J. C. and Gros-Louis, J. (1998). Chorusing and call convergence in chimpanzees: tests of three hypotheses. *Behaviour* 135: 1041–1064.

Mitani, J. C. and Nishida, T. (1993). Contexts and social correlates of long-distance calling by male chimpanzees. *Animal Behaviour* 45: 735–746.

Mitani, J. C. and Watts, D. P. (1999). Demographic influences on the hunting behavior of chimpanzees. *American Journal of Physical Anthropology* 109: 439–454.

Mitani, J. C. and Watts, D. P. (2001). Why do chimpanzees hunt and share meat? *Animal Behaviour* 61: 915–924.

Mitani, J. C. and Watts, D. P. (2005). Correlates of territorial boundary patrol behaviour in wild chimpanzees. *Animal Behaviour* 70: 1079–1086.

Mitani, J. C., Hasegawa, T., Gros-Louis, J., Marler, P., and Byrne, R. (1992). Dialects in wild chimpanzees? *American Journal of Primatology* 27: 233–243.

Mitani, J. C., Gros-Louis, J., and Macedonia, J. M. (1996). Selection for acoustic variability within the vocal repertoire of wild chimpanzees. *International Journal of Primatology* 17: 569–583.

Mitani, J. C., Hunley, K. L., and Murdoch, M. E. (1999). Geographic variation in the calls of wild chimpanzees: A reassessment. *American Journal of Primatology* 47: 133–151.

Mitani, J. C., Merriwether, D. A., and Zhang, C. (2000a). Male affiliation, cooperation and kinship in wild chimpanzees. *Animal Behaviour* 59: 885–893.

Mitani, J. C., Struhsaker, T. T., and Lwanga, J. S. (2000b). Primate community dynamics in old growth forest over 23.5 years at Ngogo, Kibale National Park, Uganda: Implications for conservation and census methods. *International Journal of Primatology* 21: 269–286.

Mitani, J. C., Watts, D. P., and Lwanga, J. S. (2002a). Ecological and social correlates of chimpanzee party size and composition. In *Behavioural Diversity in Chimpanzees and Bonobos* (C. Boesch, G. Hohmann, and L. F. Marchant, eds.), Cambridge University Press, Cambridge, pp. 102–111.

Mitani, J. C., Watts, D. P., Pepper, J. W., and Merriwether, A. (2002b). Demographic and social constraints on male chimpanzee behaviour. *Animal Behaviour* 64: 727–737.

Mitani, J. C., Watts, D. P., and Amsler, S. J. (2010). Lethal intergroup aggression leads to territorial expansion in wild chimpanzees. *Current Biology* 20: R507–R508.

Mitra Setia, T., Delgado, R. A., Utami Atmoko, S. S., Singleton, I., and van Schaik, C. P. (2009). Social organization and male–female relationships. In *Orangutans: Geographic Variation in Behavioral Ecology and Conservation* (S. A. Wich, S. S. Utami Atmoko, T. Mitra Setia, and C. P. van Schaik, eds.), Oxford University Press, Oxford, pp. 245–253.

Mjungu, D. C. (2010). *Dynamics of Intergroup Competition in Two Neighboring Chimpanzee Communities*. Ph.D. dissertation, University of Minnesota, http://conservancy.umn.edu/bitstream/95275/1/MJUNGU_umn_0130E_11279.pdf

Möbius, Y., Boesch, C., Koops, K., Matsuzawa, T., and Humle, T. (2008). Cultural differences in army ant predation by West African chimpanzees? A comparative study of microecological variables. *Animal Behaviour* 76: 37–45.

Moorjani, P., Amorim, C. E. G., Arndt, P. F., and Przeworski, M. (2016). Variation in the molecular clock of primates. *Proceedings of the National Academy of Sciences of the United States of America* 113: 10607–10612.

Muller, M. N. (2002). Agonistic relations among Kanyawara chimpanzees. In *Behavioural Diversity in Chimpanzees and Bonobos* (C. Boesch, G. Hohmann, and L. F. Marchant, eds.), Cambridge University Press, Cambridge, pp. 112–124.

Muller, M. N. and Wrangham, R. W. (2004a). Dominance, aggression and testosterone in wild chimpanzees: A test of the "challenge hypothesis." *Animal Behaviour* 67: 113–123.

Muller, M. N. and Wrangham, R. W. (2014). Mortality rates among Kanyawara chimpanzees. *Journal of Human Evolution* 67: 107–114.

Muller, M. N., Emery Thompson, M., and Wrangham, R. W. (2006). Male chimpanzees prefer mating with older females. *Current Biology* 16: 2234–2238.

Muller, M. N., Kahlenberg, S. M., Emery Thompson, M., and Wrangham, R. W. (2007). Male coercion and the costs of promiscuous mating for female chimpanzees. *Proceedings of the Royal Society B* 274: 1009–1014.

Muller, M. N., Kahlenberg, S. M., and Wrangham, R. W. (2009). Male aggression against females and sexual coercion in chimpanzees. In *Sexual Coercion in Primates and Humans* (M. N. Muller and R. W. Wrangham, eds.), Harvard University Press, Cambridge, MA, pp. 184–217.

Muller, M. N., Emery Thompson, M., Kahlenberg, S. M., and Wrangham, R. W. (2011). Sexual coercion by male chimpanzees shows that female choice may be more apparent than real. *Behavioral Ecology and Sociobiology* 65: 921–933.

Muller, M. N., Wrangham, R. W., and Pilbeam, D. R., eds. (2017). *Chimpanzees and Human Evolution*. Belknap Press of Harvard University Press, Cambridge, MA.

Murray, C. M. (2007). Method for assigning categorical rank in female *Pan troglodytes schweinfurthii* via the frequency of approaches. *International Journal of Primatology* 28: 853–864.

Murray, C. M., Eberly, L. E., and Pusey, A. E. (2006). Foraging strategies as a function of season and rank among wild female chimpanzees (*Pan troglodytes*). *Behavioral Ecology* 17(6): 1020–1028.

Murray, C. M., Mane, S. V., and Pusey, A. E. (2007a). Dominance rank influences female space use in wild chimpanzees, *Pan troglodytes*: Towards an ideal despotic distribution. *Animal Behaviour* 74: 1795–1804.

Murray, C. M., Wroblewski, E., and Pusey, A. E. (2007b). New case of intragroup infanticide in the chimpanzees of Gombe National Park. *International Journal of Primatology* 28: 23–37.

Murray, C. M., Gilby, I. C., Mane, S. V., and Pusey, A. E. (2008). Adult male chimpanzees inherit maternal ranging patterns. *Current Biology* 18: 20–24.

Murray, C. M., Lonsdorf, E. V., Eberly, L. E., and Pusey, A. E. (2009). Reproductive energetics in free-living female chimpanzees (*Pan troglodytes schweinfurthii*). *Behavioral Ecology* 20: 1211–1216.

Musgrave, S., Morgan, D., Lonsdorf, E., Mundry, R., and Sanz, C. (2016). Tool transfers are a form of teaching among chimpanzees. *Scientific Reports* 6: 34783.

Nadler, R. D., Graham, C. E., Gosselin, R. E., and Collins, D. C. (1985). Serum levels of gonadotropins and gonadal steroids, including testosterone, during the menstrual cycle of the chimpanzee (*Pan troglodytes*). *American Journal of Primatology* 9: 273–284.

Nakamura, M. (2002). Grooming-hand-clasp in Mahale M group chimpanzees: Implications for culture in social behaviors. In *Behavioural Diversity in Chimpanzees and Bonobos* (C. Boesch, G. Hohmann, and L. F. Marchant, eds.), Cambridge University Press, Cambridge, pp. 71–83.

Nakamura, M. (2010). Poke-type social scratching persists at Mahale. *Pan Africa News* 17: 15–17.

Nakamura, M. (2011). Comparison of social behaviors. In *The Chimpanzees of Bossou and Nimba* (T. Matsuzawa, T. Humle, and Y. Sugiyama, eds.), Springer, New York, NY, pp. 251–263.

Nakamura, M. and Nishida, T. (2013). Ontogeny of a social custom in wild chimpanzees: Age changes in grooming hand-clasp at Mahale. *American Journal of Primatology* 75: 186–196.

Nakamura, M. and Uehara, S. (2004). Proximate factors of different types of grooming hand-clasp in Mahale chimpanzees: Implications for chimpanzee social customs. *Current Anthropology* 45: 108–114.

Nakamura, M., Marchant, L. F., and Nishida, T. (2000). Social scratch: Another social custom in wild chimpanzees? *Primates* 41: 237–248.

Nakamura, M. *et al.* (2013). Ranging behavior of Mahale chimpanzees: A 16 year study. *Primates* 54: 171–182.

Nakamura, M., Hosaka, K., Itoh, N., and Zamma, K., eds. (2015). *Mahale Chimpanzees: 50 Years of Research.* Cambridge University Press, Cambridge.

Nakazawa, N., Hanamura, S., Inoue, E., Nakatsukasa, M., and Nakamura, M. (2013). A leopard ate a chimpanzee: First evidence from East Africa. *Journal of Human Evolution* 65: 334–337.

Nekaris, K. A. I. and Bearder, S. K. (2011). The lorisiform primates of Asia and mainland Africa. In *Primates in Perspective* (C. J. Campbell, A. Fuentes, K. C. MacKinnon, S. K. Bearder, and R. M. Stumpf, eds.), 2nd edn., Oxford University Press, Oxford, pp. 34–54.

Newton-Fisher, N. E. (1999a). Association by male chimpanzees: A social tactic? *Behaviour* 136(6): 705–730.

Newton-Fisher, N. E. (1999b). Infant killers of Budongo. *Folia Primatologica* 70: 167–169.

Newton-Fisher, N. E. (2002). Relationships of male chimpanzees in the Budongo Forest, Uganda. In *Behavioural Diversity in Chimpanzees and Bonobos* (C. Boesch, G. Hohmann, and L. F. Marchant, eds.), Cambridge University Press, Cambridge, pp. 125–137.

Newton-Fisher, N. E. (2003). The home range of the Sonso community of chimpanzees from the Budongo Forest, Uganda. *African Journal of Ecology* 41: 150–156.

Newton-Fisher, N. E. (2006). Female coalitions against male aggression in wild chimpanzees of the Budongo Forest. *International Journal of Primatology* 27: 1589–1599.

Newton-Fisher, N. E. and Lee, P. C. (2011). Grooming reciprocity in wild male chimpanzees. *Animal Behaviour* 81: 439–446.

Newton-Fisher, N. E., Reynolds, V., and Plumptre, A. J. (2000). Food supply and chimpanzee (*Pan troglodytes schweinfurthii*) party size in the Budongo Forest Reserve, Uganda. *International Journal of Primatology* 21: 613–628.

Newton-Fisher, N. E., Emery Thompson, M., Reynolds, V., Boesch, C., and Vigilant, L. (2010). Paternity and social rank in wild chimpanzees (*Pan troglodytes*) from the Budongo Forest, Uganda. *American Journal of Physical Anthropology* 142: 417–428.

Nishida, T. (1968). The social group of wild chimpanzees in the Mahali Mountains. *Primates* 9: 167–224.

Nishida, T. (1970). Social behavior and relationship among wild chimpanzees of the Mahali Mountains. *Primates* 11: 47–87.

Nishida, T. (1979). The social structure of chimpanzees of the Mahale Mountains. In *The Great Apes* (D. A. Hamburg and E. R. McCown, eds.), Benjamin/Cummings, Menlo Park, CA, pp. 73–121.

Nishida, T. (1980). The leaf-clipping display: A newly discovered expressive gesture in wild chimpanzees. *Journal of Human Evolution* 9: 117–128.

Nishida, T. (1983). Alpha status and agonistic alliance in wild chimpanzees (*Pan troglodytes schweinfurthii*). *Primates* 24(3): 318–336.

Nishida, T. (1989). Social interactions between resident and immigrant chimpanzees. In *Understanding Chimpanzees* (P. G. Heltne and L. A. Marquardt, eds.), Harvard University Press, Cambridge, MA, pp. 68–89.

Nishida, T., ed. (1990). *The Chimpanzees of the Mahale Mountains*. University of Tokyo Press, Tokyo.

Nishida, T. (1997). Sexual behavior of adult male chimpanzees of the Mahale Mountains National Park, Tanzania. *Primates* 38(4): 379–398.

Nishida, T. (2010). *Chimpanzee behavior in the wild [electronic resource]: An audiovisual encyclopedia*. Springer, Tokyo.

Nishida, T. (2012). *Chimpanzees of the Lakeshore: Natural History and Culture at Mahale*. Cambridge University Press, Cambridge.

Nishida, T. and Hiraiwa-Hasegawa, M. (1985). Responses to a stranger mother-son pair in the wild chimpanzee: A case report. *Primates* 26: 1–13.

Nishida, T. and Hosaka, K. (1996). Coalition strategies among adult male chimpanzees of the Mahale Mountains, Tanzania. In *Great Ape Societies* (W. C., McGrew, L. F. Marchant, and T. Nishida, eds.), Cambridge University Press, Cambridge, pp. 114–134.

Nishida, T. and Kawanaka, K. (1972). Inter-unit-group relationships among wild chimpanzees of the Mahali Mountains. *Kyoto University Africa Studies* 7: 131–169.

Nishida, T. and Kawanaka, K. (1985). Within-group cannibalism by adult male chimpanzees. *Primates* 26(3): 274–284.

Nishida, T. and Turner, L. A. (1996). Food transfer between mother and infant chimpanzees of the Mahale Mountains National Park, Tanzania. *International Journal of Primatology* 17: 947–968.

Nishida, T., Uehara, S., and Nyundo, R. (1979). Predatory behavior among wild chimpanzees of the Mahale Mountains. *Primates* 20: 1–20.

Nishida, T., Hiraiwa-Hasegawa, M., Hasegawa, T., and Takahata, Y. (1985). Group extinction and female transfer in wild chimpanzees in the Mahale National Park, Tanzania. *Zeitschrift für Tierpsychologie* 67: 284–301.

Nishida, T., Takasaki, H., and Takahata, Y. (1990). Demography and reproductive profiles. In *The Chimpanzees of the Mahale Mountains* (T. Nishida, ed.), University of Tokyo Press, Tokyo, pp. 63–97.

Nishida, T., Hasegawa, T., Hayaki, H., Takahata, Y., and Uehara, S. (1992). Meat-sharing as a coalition strategy by an alpha male chimpanzee? In *Topics in Primatology, Vol. 1: Human Origins* (T. Nishida, W. C. McGrew, P. Marler, M. Pickford, and F. B. M. de Waal, eds.), University of Tokyo Press, Tokyo, pp. 159–174.

Nishida, T., Hosaka, K., and Hamai, M. (1995). A within-group gang attack on a young adult male chimpanzee: Ostracism of an ill-mannered member? *Primates* 36: 207–211.

Nishida, T., Corp, N., Hamai, M., Hasegawa, T., Hiraiwa-Hasegawa, M., Hosaka, K., Hunt, K. D., Itoh, N., Kawanaka, K., Matsumoto-Oda, A., Mitani, J. C., Nakamura, M., Norikoshi, K., Sakamaki, T., Turner, L., Uehara, S., and Zamma, K. (2003). Demography, female life history, and reproductive profiles among the chimpanzees of Mahale. *American Journal of Primatology* 59: 99–121. doi:10.1002/ajp.10068

Nishida, T., Mitani, J. C., and Watts, D. P. (2004). Variable grooming behaviours in wild chimpanzees. *Folia Primatologica* 75: 31–36.

Nishida, T., Matsusaka, T., and McGrew, W. C. (2009). Emergence, propagation or disappearance of novel behavioral patterns in the habituated chimpanzees of Mahale: A review. *Primates* 50: 23–36.

Nishida, T., Zamma, K., Matsusaka, T., Inaba, A., and McGrew, W. C. (2010). *Chimpanzee Behavior in the Wild: An Audio-visual Encyclopedia*. Springer, Tokyo.

Nishie, H. (2015). Use of tools and other objects. In *Mahale Chimpanzees: 50 Years of Research* (M. Nakamura, K. Hosaka, N. Itoh, and K. Zamma, eds.), Cambridge University Press, Cambridge, pp. 568–582.

Notman, H. and Rendall, D. (2005). Contextual variation in chimpanzee pant hoots and its implications for referential communication. *Animal Behaviour* 70: 177–190.

Nunn, C. L. (1999). The evolution of exaggerated sexual swellings in primates and the graded-signal hypothesis. *Animal Behaviour* 58: 229–246.

O'Malley, R. C., Wallauer, W., Murray, C. M., and Goodall, J. (2012). The appearance and spread of ant fishing among the Kasekela chimpanzees of Gombe: A possible case of intercommunity cultural transmission. *Current Anthropology* 53: 650–663.

Ohashi, G. and Matsuzawa, T. (2011). Deactivation of snares by wild chimpanzees. *Primates* 52: 1–5.

Otali, E. and Gilchrist, J. S. (2006). Why chimpanzee (*Pan troglodytes schweinfurthii*) mothers are less gregarious than nonmothers and males: The infant safety hypothesis. *Behavioral Ecology and Sociobiology* 59: 561–570.

Packer, C. and Ruttan, L. (1988). The evolution of cooperative hunting. *American Naturalist* 132: 159–198.

Parish, A. R. (1996). Female relationships in bonobos (*Pan paniscus*). *Human Nature* 7: 61–96.

Parr, L. A. and Waller, B. M. (2006). Understanding chimpanzee facial expression: Insights into the evolution of communication. *Social Cognitive and Affective Neuroscience* 1: 221–228. doi:10.1093/scan/nsl031

Parr, L. A., Waller, B. M., Vick, S. J., and Bard, K. A. (2007). Classifying chimpanzee facial expressions using muscle action. *Emotion* 7: 172–181.

Pepper, J. W., Mitani, J. C., and Watts, D. P. (1999). General gregariousness and specific social preferences among wild chimpanzees. *International Journal of Primatology* 20: 613–632.

Perry, S. (2009). Are nonhuman primates likely to exhibit cultural capacities like those of humans? In *The Question of Animal Culture* (K. N. Laland and B. G. Galef, eds.), Harvard University Press, Cambridge, MA, pp. 247–268.

Pieta, K. (2008). Female mate preferences among *Pan troglodytes schweinfurthii* of Kanyawara, Kibale National Park, Uganda. *International Journal of Primatology* 29: 845–864.

Pika, S. and Mitani, J. C. (2006). Referential gestural communication in wild chimpanzees (*Pan troglodytes*). *Current Biology* 16(6): R191–R192.

Pintea, L., Pusey, A., Wilson, M., Gilby, I., Collins, A., Kamenya, S., and Goodall, J. (2011). Long-term ecological changes affecting the chimpanzees of Gombe National Park, Tanzania. In *The Ecological Impact of Long-Term Changes in Africa's Rift Valley* (A. J. Plumptre, ed.), Nova Science, New York, NY, pp. 227–247.

Plooij, F. X. (1978a). Tool use during chimpanzees' bushpig hunt. *Carnivore* 1: 103–106.

Plooij, F. X. (1978b). Some basic traits of language in wild chimpanzees? In *Action, Gesture and Symbol* (A. Lock, ed.), Academic Press, London, pp. 111–131.

Plooij, F. X. (1980). *The Behavioural Development of Free-living Chimpanzee Babies and Infants*. Ph.D. dissertation, Rijksuniversiteit te Groningen.

Plumptre, A. J. and Reynolds, V. (1994). The effect of selective logging on the primate populations in the Budongo Forest Reserve, Uganda. *Journal of Applied Ecology* 31: 631–641.

Plumptre, A. J. et al. (2010). *Eastern Chimpanzee* (Pan troglodytes schweinfurthii): *Status Survey and Conservation Action Plan 2010–2020*. IUCN/SSC Primate Specialist Group, Gland.

Polansky, L. and Boesch, C. (2013). Long-term changes in fruit phenology in a West African lowland tropical rain forest are not explained by rainfall. *Biotropica* 45: 434–440.

Pontzer, H. and Wrangham, R. W. (2006). Ontogeny of ranging in wild chimpanzees. *International Journal of Primatology* 27: 295–309.

Potts, K. B., Watts, D. P., and Wrangham, R. W. (2011). Comparative feeding ecology of two communities of chimpanzees (*Pan troglodytes*) in Kibale National Park, Uganda. *International Journal of Primatology* 32: 669–690.

Power, M. (1991). *The Egalitarians – Human and Chimpanzee*. Cambridge University Press, Cambridge.

Prasetyo, D., Ancrenaz, M., Morrogh-Bernard, H. C., Utami Atmoko, S. S., Wich, S. A., and van Schaik, C. P. (2009). Nest building in orangutans. In *Orangutans: Geographic Variation in Behavioral Ecology and Conservation* (S. A. Wich, S. S. Utami Atmoko, T. M. Setia, and C. P. van Schaik, eds.), Oxford University Press, Oxford, pp. 269–277.

Price, T., Wadewitz, P., Cheney, D. L., Seyfarth, R. M., Hammerschmidt, K., and Fischer, J. (2015). Vervets revisited: A quantitative analysis of alarm call structure and context specificity. *Scientific Reports* 5: 1–11, http://dx.doi.org/10.1038/srep13220

Pruetz, J. D. (2006). Feeding ecology of savanna chimpanzees (*Pan troglodytes verus*) at Fongoli, Senegal. In *Feeding Ecology in Apes and Other Primates* (G. Hohmann, M. M. Robbins, and C. Boesch, eds.), Cambridge University Press, Cambridge, pp. 161–182.

Pruetz, J. D. and Bertolani, P. (2007). Savanna chimpanzees, *Pan troglodytes verus*, hunt with tools. *Current Biology* 17: 412–417.

Pruetz, J. D., Bertolani, P., Ontl, K. B., Lindshield, S., Shelley, M., and Wessling, E. G. (2015). New evidence on the tool-assisted hunting exhibited by chimpanzees (*Pan troglodytes verus*) in a savannah habitat at Fongoli, Sénégal. *Royal Society Open Science* 2: 140507. doi:10.1098/rsos.140507

Pusey, A. E. (1980). Inbreeding avoidance in chimpanzees. *Animal Behaviour* 28: 543–552.

Pusey, A. E. (1983). Mother–offspring relationships in chimpanzees after weaning. *Animal Behaviour* 31: 363–377.

Pusey, A. E. (1990). Behavioural changes at adolescence in chimpanzees. *Behaviour* 115: 203–246.

Pusey, A. E. and Packer, C. (1987). Dispersal and philopatry. In *Primate Societies* (B. B. Smuts, D. L. Cheney, R. M. Seyfarth, R. W. Wrangham, and T. T. Struhsaker, eds.), University of Chicago Press, Chicago, IL, pp. 250–266.

Pusey, A. E. and Schroepfer-Walker, K. (2013). Female competition in chimpanzees. *Philosophical Transactions of the Royal Society B* 368: 20130077; http://dx.doi.org/10.1098/rstb.2013.0077

Pusey, A. E, Williams, J., and Goodall, J. (1997). The influence of dominance rank on the reproductive success of female chimpanzees. *Science* 277: 828–831.

Pusey, A. E. and Wolf, M. (1996). Inbreeding avoidance in animals. *Trends in Ecology and Evolution* 11: 201–206.

Pusey, A. E., Oehlert, G. W., Williams, J. M., and Goodall, J. (2005). Influence of ecological and social factors on body mass of wild chimpanzees. *International Journal of Primatology* 26(1): 3–31.

Pusey, A. E., Pintea, L., Wilson, M. L., Kamenya, S., and Goodall, J. (2007). The contribution of long-term research at Gombe National Park to chimpanzee conservation. *Conservation Biology* 21: 623–634.

Pusey, A. E., Murray, C., Wallauer, W., Wilson, M., Wroblewski, E., and Goodall, J. (2008a). Severe aggression among female *Pan troglodytes schweinfurthii* at Gombe National Park, Tanzania. *International Journal of Primatology* 29: 949–973.

Pusey, A. E., Wilson, M. L., and Collins, D. A. (2008b). Human impacts, disease risk, and population dynamics in the chimpanzees of Gombe National Park, Tanzania. *American Journal of Primatology* 70: 738–744.

Reader, S. M. and Laland, K. N. (2002). Social intelligence, innovation, and enhanced brain size in primates. *Proceedings of the National Academy of Sciences of the U.S.A.* 99: 436–441.

Reader, S. M., Hager, Y., and Laland, K. N. (2011). The evolution of primate general intelligence and cultural intelligence. *Proceedings of the Royal Society B* 366: 1017–1027.

Reichard, U. H. and Barelli, C. (2008). Life history and reproductive strategies of Khao Yai *Hylobates lar*: Implications for social evolution in apes. *International Journal of Primatology* 29: 823–844.

Reynolds, V. (2005). *The Chimpanzees of the Budongo Forest: Ecology, Behaviour, and Conservation*. Oxford University Press, Oxford.

Reynolds, V. and Reynolds, F. (1965). Chimpanzees of the Budongo Forest. In *Primate Behavior: Field Studies of Monkeys and Apes* (I. DeVore, ed.), Holt, Rinehart & Winston, New York, NY, pp. 368–424.

Riede, T., Owren, M. J., and Arcadi, A. Clark (2004). Nonlinear acoustics in pant hoots of common chimpanzees (*Pan troglodytes*): Frequency jumps, subharmonics, biphonation, and deterministic chaos. *American Journal of Primatology* 64: 277–291.

Riede, T., Arcadi, A. Clark, and Owren, M. J. (2007). Nonlinear acoustics in pant hoots of common chimpanzees (*Pan troglodytes*): Vocalizing at the edge. *Journal of the Acoustical Society of America* 121: 1758–1767.

Riedel, J., Franz, M., and Boesch, C. (2011). How feeding competition determines female gregariousness and ranging in the Taï National Park, Côte d'Ivoire. *American Journal of Primatology* 73: 305–313.

Robbins, M. M. (2011). Gorillas: Diversity in ecology and behavior. In *Primates in Perspective* (C. J. Campbell, A. Fuentes, K. C. MacKinnon, S. K. Bearder, and R. M. Stumpf, eds.), 2nd edn., Oxford University Press, Oxford, pp. 326–339.

Roberts, A. I., Vick, S., and Buchanan-Smith, H. M. (2013). Communicative intentions in wild chimpanzees: persistence and elaboration in gestural signalling. *Animal Cognition* 16: 187–196.

Roberts, A. I., Roberts, S. G. B., and Vick, S. (2014). The repertoire and intentionality of gestural communication in wild chimpanzees. *Animal Cognition* 17: 317–336.

Russon, A. E. and Begun, D. R., eds. (2004). *The Evolution of Thought*. Cambridge University Press, Cambridge.

Sakamaki, T. (2011). Submissive pant-grunt greetings of female chimpanzees in Mahale Mountains National Park, Tanzania. *African Study Monographs* 32: 25–41.

Sakamaki, T. and Hayaki, H. (2015). Greetings and dominance. In *Mahale Chimpanzees: 50 Years of Research* (M. Nakamura, K. Hosaka, N. Itoh, and K. Zamma, eds.), Cambridge University Press, Cambridge, pp. 459–471.

Sakamaki, T., Itoh, N., and Nishida, T. (2001). An attempted within-group infanticide in wild chimpanzees. *Primates* 42: 359–366.

Sakura, O. and Matsuzawa, T. (1991). Flexibility of nut-cracking behavior using stone hammers and anvils of wild chimpanzees: An experimental analysis. *Ethology* 87: 237–248.

Samuni, L., Mundry, R., Terkel, J., Zuberbühler, K., and Hobaiter, C. (2014). Socially learned habituation to human observers in wild chimpanzees. *Animal Cognition* 17: 997–1005.

Sandel, A. A., Reddy, R. B., and Mitani, J. C. (2017). Adolescent male chimpanzees do not form a dominance hierarchy with their peers. *Primates* 58: 39–49.

Sanz, C. M. and Morgan, D. B. (2009). Flexible and persistent tool-using strategies in honey-gathering by wild chimpanzees. *International Journal of Primatology* 30: 411–427.

Sanz, C. M. and Morgan, D. B. (2010). The complexity of chimpanzee tool-use behaviors. In *The Mind of the Chimpanzee* (E. V. Lonsdorf, S. R. Ross, and T. Matsuzawa, eds.), University of Chicago Press, Chicago, IL, pp. 127–140.

Sanz, C. M., Deblauwe, I., Tagg, N., and Morgan, D. B. (2014). Insect prey characteristics affecting regional variation in chimpanzee tool use. *Journal of Human Evolution* 71: 28–37.

Scally, A. *et al.* (2012). Insights into hominid evolution from the gorilla genome sequence. *Nature* 483: 169–175.

Scarantino, A. and Clay, Z. (2015). Contextually variable signals can be functionally referential. *Animal Behaviour* 100: http://dx.doi.org/10.1016/j.anbehav.2014.08.017

Schel, A. M., Machanda, Z., Townsend, S. W., Zuberbühler, K., and Slocombe, K. E. (2013). Chimpanzee food calls are directed at specific individuals. *Animal Behaviour* 86: 955–965.

Schel, A. M., Townsend, S. W., Machanda, Z., Zuberbühler, K., and Slocombe, K. E. (2014). Chimpanzee alarm call production meets key criteria for intentionality. *PLoS One* 8(10): e76674. doi:10.1371/journal.pone.0076674

Schino, G., Scucchi, S., Maestripieri, D., and Turillazzi, P. G. (1988). Allogrooming as a tension-reduction mechanism: A behavioral approach. *American Journal of Primatology* 16: 43–50.

Schöning, C., Humle, T., Möbius, Y., and McGrew, W. C. (2008). The nature of culture: Technological variation in chimpanzee predation on army ants revisited. *Journal of Human Evolution* 55: 48–59.

Scott, J. P. and Fredericson, E. (1951). The causes of fighting in mice and rats. *Physiological Zoölogy* 24: 273–309.

Seyfarth, R. M. and Cheney, D. L. (2003). The structure of social knowledge in monkeys. In *Animal Social Intelligence* (F. B. M. de Waal and P. L. Tyack, eds.), Harvard University Press, Cambridge, MA, pp. 207–229.

Seyfarth, R. M., Cheney, D. L., and Marler, P. (1980). Vervet monkey alarm calls: Semantic communication in a free-ranging primate. *Animal Behaviour* 28: 1070–1094.

Sherrow, H. M. (2012). Adolescent male chimpanzees at Ngogo, Kibale National Park, Uganda have decided dominance relationships. *Folia Primatologica* 83: 67–75.

Shimada, M. K. (2002). Social scratch among the chimpanzees of Gombe. *Pan Africa News* 9: 21–23.

Shimada, M. K., Hayakawa, S., Fujita, S., Sugiyama, Y., and Saitou, N. (2009). Skewed matrilineal genetic composition in a small wild chimpanzee community. *Folia Primatologica* 80: 19–32.

Short, R. V. (1979). Sexual selection and its component parts, somatic and genital selection, as illustrated by man and the great apes. *Advances in the Study of Behavior* 9: 131–158.

Silk, J. B. (2012). The adaptive value of sociality. In *The Evolution of Primate Societies* (J. C. Mitani, J. Call, P. M. Kappeler, R. A. Palombit, and J. B. Silk, eds.), University of Chicago Press, Chicago, IL, pp. 552–564.

Simpson, M. J. A. (1973). The social grooming of male chimpanzees. In *Comparative Ecology and Behaviour of Primates* (R. P. Michael and J. H. Crook, eds.), Academic Press, New York, NY, pp. 41–50.

Singleton, I., Knott, C. D., Morrogh-Bernard, H. C., Wich, S. A., and van Schaik, C. P. (2009). Ranging behavior of orangutan females and social organization. In *Orangutans: Geographic Variation in Behavioral Ecology and Conservation* (S. A. Wich, S. S. Utami Atmoko, T. M. Setia, and C. P. van Schaik, eds.), Oxford University Press, Oxford, pp. 205–213.

Slater, K., Cameron, E., Turner, T., and du Toit, J. T. (2008). The influence of oestrous swellings on the grooming behaviour of chimpanzees of the Budongo Forest, Uganda. *Behaviour* 145: 1235–1246.

Slocombe, K. E. and Newton-Fisher, N. E. (2005). Food sharing between wild adult chimpanzees (*Pan troglodytes schweinfurthii*): A socially significant event? *American Journal of Primatology* 65: 385–391.

Slocombe, K. E. and Zuberbühler, K. (2005a). Agonistic screams in wild chimpanzees (*Pan troglodytes schweinfurthii*) vary as a function of social role. *Journal of Comparative Psychology* 119: 67–77.

Slocombe, K. E. and Zuberbühler, K. (2005b). Functionally referential communication in a chimpanzee. *Current Biology* 15: 1779–1784.

Slocombe, K. E. and Zuberbühler, K. (2006). Food-associated calls in chimpanzees: responses to food types or food preferences? *Animal Behaviour* 72: 989–999.

Slocombe, K. E. and Zuberbühler, K. (2007). Chimpanzees modify recruitment screams as a function of audience composition. *Proceedings of the National Academy of Sciences of the U.S.A.* 104: 17228–17233.

Slocombe, K. E., Kaller, T., Turman, L., Papworth, S., Townsend, S., and Zuberbühler, K. (2010). Production of food-associated calls in wild male chimpanzees is dependent on the composition of the audience. *Behavioral Ecology and Sociobiology* 64: 1959–1966.

Smaers, J. B. and Soligo, C. (2013). Brain reorganization, not relative brain size, primarily characterizes anthropoid brain evolution. *Proceedings of the Royal Society B* 280: 20130269. http://dx.doi.org/10.1098/rspb.2013.0269

Smith, W. J. (1965). Message, meaning, and context in ethology. *American Naturalist* 99: 405–409.

Smith, W. J. (1969). Messages in vertebrate communication. *Science* 165: 145–150.

Smith, W. J. (1977). *The Behavior of Communicating*. Harvard University Press, Cambridge, MA.

Smuts, B. B. (1986). Gender, aggression, and influence. In *Primate Societies* (B. B. Smuts, D. L. Cheney, R. M. Seyfarth, R. W. Wrangham, and T. T. Struhsaker, eds.), University of Chicago Press, Chicago, IL, pp. 400–412.

Sobolewski, M. E., Brown, J. L., and Mitani, J. C. (2013). Female parity, male aggression, and the Challenge Hypothesis. *Primates* 54: 81–88.

Stanford, C. (1995). The influence of chimpanzee predation on group size and anti-predator behaviour in red colobus monkeys. *Animal Behaviour* 49: 577–587.

Stanford, C. (1996). The hunting ecology of wild chimpanzees: Implications for the evolutionary ecology of Pliocene hominids. *American Anthropologist* 98: 96–113.

Stanford, C. (1998). *Chimpanzee and Red Colobus*. Harvard University Press, Cambridge, MA.

Stanford, C. B., Wallis, J., Mpongo, E., and Goodall, J. (1994a). Hunting decisions in wild chimpanzees. *Behaviour* 131: 1–18.

Stanford, C. B., Wallis, J., Matama, H., and Goodall, J. (1994b). Patterns of predation by chimpanzees on red colobus monkeys in Gombe National Park, 1982–1991. *American Journal of Physical Anthropology* 94: 213–228.

Steklis, H. D. and Kling, A. (1985). Neurobiology of affiliative behavior in nonhuman primates. In *The Psychobiology of Attachment and Separation* (M. Reite and T. Field, eds.), Academic Press, Orlando, FL, pp. 93–134.

Strier, K. B. (2011). *Primate Behavioral Ecology*, 4th edn., Allyn & Bacon, Needham Heights, MA.

Struhsaker, T. T. (1975). *The Red Colobus Monkey*. University of Chicago Press, Chicago, IL.

Struhsaker, T. T. (1999). Primate communities in Africa: The consequence of long-term evolution or the artifact of recent human hunting? In *Primate Communities* (J. G. Fleagle, C. Janson, and K. E. Reed, eds.), Cambridge University Press, Cambridge, pp. 289–294.

Struhsaker, T. T. (2010). *The Red Colobus Monkeys: Variation in Demography, Behavior, and Ecology of Endangered Species*. Oxford University Press, Oxford.

Stumpf, R. M. (2011). Chimpanzees and bonobos: Inter- and intraspecies diversity. In *Primates in Perspective* (C. J. Campbell, A. Fuentes, K. C. MacKinnon, S. K. Bearder, and R. M. Stumpf, eds.), 2nd edn., Oxford University Press, Oxford, pp. 340–356.

Stumpf, R. M. and Boesch, C. (2005). Does promiscuous mating preclude female choice? Female sexual strategies in chimpanzees (*Pan troglodytes verus*) of the Taï National Park, Côte d'Ivoire. *Behavioral Ecology and Sociobiology* 57: 511–524.

Stumpf, R. M. and Boesch, C. (2006). The efficacy of female choice in chimpanzees of the Taï Forest, Côte d'Ivoire. *Behavioral Ecology and Sociobiology* 60: 749–765.

Stumpf, R. M. and Boesch, C. (2010). Male aggression and sexual coercion in wild West African chimpanzees, *Pan troglodytes verus*. *Animal Behaviour* 79: 333–342.

Stumpf, R. M., Emery Thompson, M., Muller, M. N., and Wrangham, R. W. (2009). The context of female dispersal in Kanyawara chimpanzees. *Behaviour* 146: 629–656.

Sugiyama, Y. (1969). Social behavior of chimpanzees in the Budongo Forest, Uganda. *Primates* 10: 197–225.

Sugiyama, Y. (1988). Grooming interactions among adult chimpanzees at Bossou, Guinea, with special reference to social structure. *International Journal of Primatology* 9: 393–407.

Sugiyama, Y. (1999). Socioecological factors of male migration at Bossou, Guinea. *Primates* 40: 61–68.

Sugiyama, Y. (2004). Demographic parameters and life history of chimpanzees at Bossou, Guinea. *American Journal of Physical Anthropology* 124: 154–165.

Sugiyama, Y. and Fujita, S. (2011). The demography and reproductive parameters of Bossou chimpanzees. In *The Chimpanzees of Bossou and Nimba* (T. Matsuzawa, T. Humle, and Y. Sugiyama, eds.), Springer, Tokyo, pp. 23–34.

Sugiyama, Y. and Koman, J. (1979). Social structure and dynamics of wild chimpanzees at Bossou, Guinea. *Primates* 20: 323–339.

Sugiyama, Y. and Koman, J. (1987). A preliminary list of chimpanzees' alimentation at Bossou, Guinea. *Primates* 28: 133–147.

Suzuki, A. (1971). Carnivority and cannibalism observed among forest-living chimpanzees. *Journal of the Anthropological Society of Nippon* 79: 30–48.

Takahata, Y. (1985). Adult male chimpanzees kill and eat a male newborn infant: Newly observed intragroup infanticide and cannibalism in Mahale National Park, Tanzania. *Folia Primatologica* 44: 161–170.

Takahata, Y. (1990a). Adult males' social relations with adult females. In *The Chimpanzees of the Mahale Mountains* (T. Nishida, ed.). University of Tokyo Press, Tokyo, pp. 133–148.

Takahata, Y. (1990b). Social relationships among adult males. In *The Chimpanzees of the Mahale Mountains* (T. Nishida, ed.). University of Tokyo Press, Tokyo, pp. 149–170.

Takahata, Y., Hasegawa, T., and Nishida, T. (1984). Chimpanzee predation in the Mahale Mountains from August 1979 to May 1982. *International Journal of Primatology* 5: 213–233.

Takasaki, H. (1985). Female life history and mating patterns among M group chimpanzees of the Mahale National Park, Tanzania. *Primates* 26(2): 121–129.

Teelen, S. (2008). Influence of chimpanzee predation on the red colobus population at Ngogo, Kibale National Park, Uganda. *Primates* 49: 41–49.

Teleki, G. (1973). *The Predatory Behavior of Wild Chimpanzees*. Bucknell University Press, Lewisburg, PA.

Tennie, C., Gilby, I.C., and Mundry, R. (2009a). The meat-scrap hypothesis: Small quantities of meat may promote cooperative hunting in wild chimpanzees (*Pan troglodytes*). *Behavioral Ecology and Sociobiology* 63: 421–431.

Tennie, C., Call, J., and Tomasello, M. (2009b). Ratcheting up the ratchet: On the evolution of cumulative culture. *Philosophical Transactions of the Royal Society B* 364: 2405–2415.

Tennie, C., Call, J., and Tomasello, M. (2012). Untrained chimpanzees (*Pan troglodytes schweinfurthii*) fail to imitate novel actions. *PLoS ONE* 7: e41548.

Tomasello, M. (1994). The question of chimpanzee culture. In *Chimpanzee Cultures* (R. W. Wrangham, W. C. McGrew, F. B. M. de Waal, and P. G. Heltne, eds.), Harvard University Press, Cambridge, MA, pp. 301–317.

Tomasello, M. (2009). The question of chimpanzee culture, plus postscript (chimpanzee culture, 2009). In *The Question of Animal Culture* (K. N. Laland and B. G. Galef, eds.), Harvard University Press, Cambridge, MA, pp. 198–221.

Tomasello, M. and Call, J. (1997). *Primate Cognition*. Oxford University Press, Oxford.

Tomasello, M. and Call, J. (2007). Introduction: Intentional communication in nonhuman primates. In *The Gestural Communication of Apes and Monkeys* (J. Call and M. Tomasello, eds.), Lawrence Erlbaum Associates, Mahwah, NJ, pp. 1–15.

Tomasello, M. and Call, J. (2010). Chimpanzee social cognition. In *The Mind of the Chimpanzee* (E. V. Lonsdorf, S. R. Ross, and T. Matsuzawa, eds.), University of Chicago Press, Chicago, IL, pp. 235–250.

Tomasello, M., Call, J., Nagell, K., Olguin, R., and Carpenter, M. (1994). The learning and use of gestural signals by young chimpanzees: A trans-generational study. *Primates* 35: 137–154.

Townsend, S. W. and Manser, M. B. (2013). Functionally referential communication in mammals: The past, present and the future. *Ethology* 119: 1–11.

Townsend, S. W., Slocombe, K. E., Emery Thompson, M., and Zuberbühler, K. (2007). Female-led infanticide in wild chimpanzees. *Current Biology* 17: 355–356.

Townsend, S. W., Deschner, T., and Zuberbühler, K. (2008). Female chimpanzees use copulation calls flexibly to prevent social competition. *PloS ONE* 3(6), e2431.

Townsend, S. W., Deschner, T., and Zuberbühler, K. (2011). Copulation calls in female chimpanzees (*Pan troglodytes schweinfurthii*) convey identity but do not accurately reflect fertility. *International Journal of Primatology* 32: 914–923.

Tutin, C. E. G. (1979). Mating patterns and reproductive strategies in a community of wild chimpanzees (*Pan troglodytes schweinfurthii*). *Behavioral Ecology and Sociobiology* 6: 29–38.

Tutin, C. E. G. and McGinnis, P. R. (1981). Chimpanzee reproduction in the wild. In *Reproductive Biology of the Great Apes* (C. E. Graham, ed.), Academic Press, New York, NY, pp. 239–264.

Tutin, C. E. G. and McGrew, W. C. (1973). Sexual behaviour of group-living adolescent chimpanzees. *American Journal of Physical Anthropology* 38: 195–199.

Uehara, S. (1986). Sex and group differences in feeding on animals by wild chimpanzees in the Mahale Mountains National Park, Tanzania. *Primates* 27: 1–13.

Uehara, S. (1997). Predation on mammals by the chimpanzee. *Primates* 38: 193–214.

Uehara, S., Nishida, T., Hamai, M., Hayaki, H., Huffman, M., Kawanaka, K., Kobayashi, S., Mitani, J., Takahata, Y., Takasaki, H., and Tsukahara, T. (1992). Characteristics of predation by the chimpanzees in the Mahale Mountains National Park, Tanzania. In *Topics in Primatology, Vol. 1: Human Origins* (T. Nishida, W. C. McGrew, P. Marler, M. Pickford, and F. B. M. de Waal, eds.), University of Tokyo Press, Tokyo, pp. 143–158.

Utami Atmoko, S. S., Singleton, I., van Noordwijk, M. A., van Schaik, C. P., and Setia, T. M. (2009). Male–male relationships in orangutans. In *Orangutans: Geographic Variation in Behavioral Ecology and Conservation* (S. A. Wich, S. S. Utami Atmoko, T. M. Setia, and C. P. van Schaik, eds.), Oxford University Press, Oxford, pp. 225–233.

Vahed, K. and Parker, D. J. (2012). The evolution of large testes: Sperm competition or male mating rate? *Ethology* 118: 107–117.

van Hooff, J. A. R. A. M. (1967). The facial displays of the catarrhine monkeys and apes. In *Primate Ethology* (D. Morris, ed.), Weidenfeld and Nicolson, London, pp. 7–68.

van Leeuwen, E. J. C. and Haun, D. B. M. (2014). Conformity without majority? The case for demarcating social from majority influences. *Animal Behaviour* 96: 187–194.

van Leeuwen, E. J. C., Cronin, K. A., Haun, D. B. M., Mundry, R., and Bodamer, M. D. (2012). Neighbouring chimpanzee communities show different preferences in social grooming behaviour. *Proceedings of the Royal Society B* 279: 4362–4367.

van Leeuwen, E. J. C., Cronin, K. A., Schütte, S., Call, J., and Haun, D. B. M. (2013). Chimpanzees (*Pan troglodytes*) flexibly adjust their behaviour in order to maximize payoffs, not to conform to majorities. *PLoS One* 8: e80945.

van Leeuwen, E. J. C., Cronin, K. A., Mundry, R., Bodamer, M., and Haun, D. B. M. (2017). Chimpanzee culture extends beyond matrilineal family units. *Current Biology* 27: R588–R590.

van Schaik, C. P., van Noordwijk, M. A., and Nunn, C. L. (1999). Sex and social evolution in primates. In *Comparative Primate Socioecology* (P. C. Lee, ed.), Cambridge University Press, Cambridge, pp. 204–231.

Vasey, N. (2006). Impact of seasonality and reproduction on social structure, ranging patterns, and fission-fusion social organization in red ruffed lemurs. In *Lemurs: Ecology and Adaptation* (L. Gould and M. Sauther, eds.), Springer, New York, NY, pp. 275–303.

Visalberghi, E. and Fragaszy, D. (2006). What is challenging about tool use? The capuchin's perspective. In *Comparative Cognition: Experimental Explorations of Animal Intelligence* (E. A. Wasserman and T. R. Zentall, eds.), Oxford University Press, Oxford, pp. 529–552.

Wakefield, M. L. (2008). Grouping patterns and competition among female *Pan troglodytes schweinfurthii* at Ngogo, Kibale National Park, Uganda. *International Journal of Primatology* 29: 907–929.

Wakefield, M. L. (2013). Social dynamics among females and their influence on social structure in an East African chimpanzee community. *Animal Behaviour* 85: 1303–1313.

Waller, B. M. and Dunbar, R. I. M. (2005). Differential behavioural effects of silent bared teeth display and relaxed open mouth display in chimpanzees (*Pan troglodytes*). *Ethology* 111: 129–142.

Wallis, J. (1992). Chimpanzee genital swelling and its role in the pattern of sociosexual behavior. *American Journal of Primatology* 28: 101–113.

Wallis, J. (1997). A survey of reproductive parameters in the free-ranging chimpanzees of Gombe National Park. *Journal of Reproduction and Fertility* 109: 297–307.

Watson, S. K., Townsend, S. W., Schel, A. M., Wilke, C., Wallace, E. K., Cheng, L., West, V., and Slocombe, K. E. (2015a). Vocal learning in the functionally referential food grunts of chimpanzees. *Current Biology* 25: 495–499.

Watson, S. K., Townsend, S. W., Schel, A. M., Wilke, C., Wallace, E. K., Cheng, L., West, V., and Slocombe, K. E. (2015b). Reply to Fisher et al. *Current Biology* 25: R1030–R1031.

Watts, D. P. (1989). Infanticide in mountain gorillas: New cases and a reconsideration of the evidence. *Ethology* 81: 1–18.

Watts, D. P. (1998). Coalitionary mate guarding by male chimpanzees at Ngogo, Kibale National Park, Uganda. *Behavioral Ecology and Sociobiology* 44: 43–55.

Watts, D. P. (2000a). Grooming between male chimpanzees at Ngogo, Kibale National Park. I. Partner number and diversity and grooming reciprocity. *International Journal of Primatology* 21: 189–210.

Watts, D. P. (2000b). Grooming between male chimpanzees at Ngogo, Kibale National Park. II. Influence of male rank and possible competition for partners. *International Journal of Primatology* 21: 211–238.

Watts, D. P. (2002). Reciprocity and interchange in the social relationships of wild male chimpanzees. *Behaviour* 139: 343–370.

Watts, D. P. (2006). Conflict resolution in chimpanzees and the valuable-relationships hypothesis. *International Journal of Primatology* 27: 1337–1364.

Watts, D. P. (2007). Effects of male group size, parity, and cycle stage on female chimpanzee copulation rates at Ngogo, Kibale National Park, Uganda. *Primates* 48: 222–231.

Watts, D. P. (2008). Scavenging by chimpanzees at Ngogo and the relevance of chimpanzee scavenging to early hominin behavioral ecology. *Journal of Human Evolution* 54: 125–133.

Watts, D. P. (2012). Long-term research on chimpanzee behavioral ecology in Kibale National Park, Uganda. In *Long-Term Field Studies of Primates* (P. M. Kappeler and D.P. Watts, eds.), Springer-Verlag, Berlin, pp. 313–338.

Watts, D. P. (2015). Mating behavior of adolescent male chimpanzees (*Pan troglodytes*) at Ngogo, Kibale National Park, Uganda. *Primates* 56: 163–172.

Watts, D. P. and Amsler, S. J. (2013). Chimpanzee-red colobus encounter rates show a red colobus population decline associated with predation by chimpanzees at Ngogo. *American Journal of Primatology* 75: 927–937.

Watts, D. P. and Mitani, J. C. (2001). Boundary patrols and intergroup encounters in wild chimpanzees. *Behaviour* 138: 299–327.

Watts, D. P. and Mitani, J. C. (2002a). Hunting behavior of chimpanzees at Ngogo, Kibale National Park, Uganda. *International Journal of Primatology* 23: 1–28.

Watts, D. P. and Mitani, J. C. (2002b). Hunting and meat sharing by chimpanzees at Ngogo, Kibale National Park, Uganda. In *Behavioural Diversity in Chimpanzees and Bonobos* (C. Boesch, G. Hohmann, and L. F. Marchant, eds.), Cambridge University Press, Cambridge, pp. 244–255.

Watts, D. P. and Mitani, J. C. (2015). Hunting and prey switching by chimpanzees (*Pan troglodytes schweinfurthii*) at Ngogo. *International Journal of Primatology* 36: 728–748.

Watts, D. P., Colmenares, F., and Arnold, K. (2000). Redirection, consolation, and male policing. In *Natural Conflict Resolution* (F. Aureli and F. B. M. de Waal, eds.), University of California Press, Berkeley, CA, pp. 281–301.

Watts, D. P., Mitani, J. C., and Sherrow, H. M. (2002). New cases of inter-community infanticide by male chimpanzees at Ngogo, Kibale National Park, Uganda. *Primates* 43: 263–270.

Watts, D. P., Potts, K. B., Lwanga, J. S., and Mitani, J. C. (2012a). Diet of chimpanzees (*Pan troglodytes schweinfurthii*) at Ngogo, Kibale National Park, Uganda, 1. Diet composition and diversity. *American Journal of Primatology* 74: 114–129. doi:10.1002/ajp.21016

Watts, D. P., Potts, K. B., Lwanga, J. S., and Mitani, J. C. (2012b). Diet of chimpanzees (*Pan troglodytes schweinfurthii*) at Ngogo, Kibale National Park, Uganda, 2. Temporal variation and fallback foods. *American Journal of Primatology* 74: 130–144.

Webster, T.H., Hodson, P. R., and Hunt, K. D. (2009). Grooming hand-clasp by chimpanzees of the Mugiri community, Toro-Semliki Wildlife Reserve, Uganda. *Pan African News* 16: 5–7.

Wheeler, B. C. and Fischer, J. (2012). Functionally referential signals: A promising paradigm whose time has passed. *Evolutionary Anthropology* 21: 195–205.

Wheeler, B. C. and Fischer, J. (2015). The blurred boundaries of functional reference: A response to Scarantino & Clay. *Animal Behaviour* 100, http://dx .doi.org/10.1016/j.anbehav.2014.11.007

White, F. J. and Wood, K. D. (2007). Female feeding priority in bonobos, *Pan paniscus*, and the question of female dominance. *American Journal of Primatology* 69: 837–850.

Whiten, A. (2010). A coming of age for cultural panthropology. In *The Mind of the Chimpanzee* (E. V. Lonsdorf, S. R. Ross, and T. Matsuzawa, eds.), University of Chicago Press, Chicago, IL, pp. 87–100.

Whiten, A. (2012). Social learning, traditions, and culture. In *The Evolution of Primate Societies* (J. C. Mitani, J. Call, P. M. Kappeler, R. A. Palombit, and J. B. Silk, eds.), University of Chicago Press, Chicago, IL, pp. 682–700.

Whiten, A. (2015). Experimental studies illuminate the cultural transmission of percussive technologies in *Homo* and *Pan*. *Philosophical Transactions of the Royal Society B* 370: 20140359.

Whiten, A., Goodall, J., McGrew, W., Nishida, T., Reynolds, V., Sugiyama, Y., Tutin, C., Wrangham, R., and Boesch, C. (1999). Cultures in chimpanzees. *Nature* 399: 682–685.

Whiten, A., Goodall, J., McGrew, W., Nishida, T., Reynolds, V., Sugiyama, Y., Tutin, C., Wrangham, R., and Boesch, C. (2001). Charting cultural variation in chimpanzees. *Behaviour* 138(11): 1481–1516.

Whiten, A., Horner, V., and de Waal, F. B. M. (2005). Conformity to cultural norms of tool use in chimpanzees. *Nature* 437: 737–740.

Whiten, A., Spiteri, A., Horner, V., Bonnie, K. E., Lambeth, S. P., Schapiro, S. J., and de Waal, F. B. M. (2007). Transmission of multiple traditions within and between chimpanzee groups. *Current Biology* 17: 1038–1043.

Whiten, A., Caldwell, C. A., and Mesoudi, A. (2016). Cultural diffusion in humans and other animals. *Current Opinion in Psychology* 8: 15–21.

Williams, J. M., Pusey, A. E., Carlis, J. V., Farm, B. P., and Goodall, J. (2002a). Female competition and male territorial behavior influence female chimpanzees' ranging patterns. *Animal Behaviour* 63: 347–360.

Williams, J. M., Liu, H., and Pusey, A. E. (2002b). Costs and benefits of grouping for female chimpanzees at Gombe. In *Behavioural Diversity in Chimpanzees and Bonobos* (C. Boesch, G. Hohmann, and L. F. Marchant, eds.), Cambridge University Press, Cambridge, pp. 192–203.

Williams, J. M., Oehlert, G. W., Carlis, J. V., and Pusey, A. E. (2004). Why do male chimpanzees defend a group range? *Animal Behaviour* 68: 523–532.

Williams, J. M., Lonsdorf, E. V., Wilson, M. L., Schumacher-Stankey, J., Goodall, J., and Pusey, A. E. (2008). Causes of death in the Kasekela chimpanzees of Gombe National Park, Tanzania. *American Journal of Primatology* 70: 766–777.

Wilson, M. L. (2012). Long-term studies of the chimpanzees of Gombe National Park, Tanzania. In *Long-Term Field Studies of Primates* (P. M. Kappeler and D. P. Watts, eds.), Springer-Verlag, Berlin, pp. 357–384.

Wilson, M. L. and Wrangham, R. W. (2003). Intergroup relations in chimpanzees. *Annual Review of Anthropology* 32: 363–392.

Wilson, M. L., Hauser, M. D., and Wrangham, R. W. (2001). Does participation in intergroup conflict depend on numerical assessment, range location, or rank for wild chimpanzees? *Animal Behaviour* 61: 1203–1216.

Wilson, M. L., Wallauer, W. R., and Pusey, A. E. (2004). New cases of intergroup violence among chimpanzees in Gombe National Park, Tanzania. *International Journal of Primatology* 25: 523–549.

Wilson, M. L., Hauser, M. D., and Wrangham, R. W. (2007). Chimpanzees (*Pan troglodytes*) modify grouping and vocal behaviour in response to location-specific risk. *Behaviour* 144: 1621–1653.

Wilson, M. L., Kahlenberg, S. M., Wells, M., and Wrangham, R. W. (2012). Ecological and social factors affect the occurrence and outcomes of intergroup encounters in chimpanzees. *Animal Behaviour* 83: 277–291.

Wilson, M. L. *et al.* (2014a). Lethal aggression in *Pan* is better explained by adaptive strategies than human impacts. *Nature* 513: 414–417

Wilson, M. L. *et al.* (2014b). Human impacts are neither necessary nor sufficient to explain chimpanzee violence (or bonobo non-violence). Response to Ferguson (2014) at https://blogs.scientificamerican.com/cross-check/chimp-violence-researchers-respond-to-criticism-on-cross-check/

Wittig, R. M. (2010). The function and cognitive underpinnings of post-conflict affiliation in wild chimpanzees. In *The Mind of the Chimpanzee* (E. V. Lonsdorf, S. R. Ross, and T. Matsuzawa, eds.), University of Chicago Press, Chicago, IL, pp. 208–219.

Wittig, R. M. and Boesch, C. (2003). Food competition and linear dominance hierarchy among female chimpanzees of the Taï National Park. *International Journal of Primatology* 24: 847–867.

Wittig, R. M. and Boesch, C. (2005). How to repair relationships – Reconciliation in wild chimpanzees (*Pan troglodytes*). *Ethology* 111: 736–763.

Wittig, R. M., Crockford, C., Lehmann, J., Whitten, P. L., Seyfarth, R. M., and Cheney, D. L. (2008). Focused grooming networks and stress alleviation in wild female baboons. *Hormones and Behavior* 54: 170–177.

Wittig, R. M., Crockford, C., Deschner, T., Langergraber, K. E., Ziegler, T. E., and Zuberbühler, K. (2014a). Food sharing is linked to urinary oxytocin levels and bonding in related and unrelated wild chimpanzees. *Proceedings of the Royal Society B* 281: 20133096. http://dx.doi.org/10.1098/rspb.2013.3096

Wittig, R. M., Crockford, C., Langergraber, K. E., and Zuberbühler, K. (2014b). Triadic social interactions operate across time: A field experiment with wild chimpanzees. *Proceedings of the Royal Society B* 281: 2013155. http://dx.doi.org/10.1098/rspb.2013.3155

Wittiger, L. and Boesch, C. (2013). Female gregariousnesss in western chimpanzees (*Pan troglodytes verus*) is influenced by resource aggregation and the number of estrous females. *Behavioral Ecology and Sociobiology* 67: 1097–1111.

Wood, B. (2010). Reconstructing human evolution: Achievements, challenges, and opportunities. *Proceedings of the National Academy of Sciences* 107: 8902–8909.

Wood, B. M., Watts, D. P., Mitani, J. C., and Langergraber, K. E. (2017). Favorable ecological circumstances promote life expectancy in chimpanzees similar to that of human hunter-gatherers. *Journal of Human Evolution* 105: 41–56.

Wrangham, R. W. (1974). Artificial feeding of chimpanzees and baboons in their natural habitat. *Animal Behaviour* 22: 83–93.

Wrangham, R. W. (1975). *The Behavioural Ecology of Chimpanzees in Gombe National Park*. Ph.D. dissertation, University of Cambridge.

Wrangham, R. W. (1977). Feeding behaviour of chimpanzees in Gombe National Park, Tanzania. In *Primate Ecology* (T. H. Clutton-Brock, ed.), Academic Press, New York, NY, pp. 503–538.

Wrangham, R. W. (1979). Sex differences in chimpanzee dispersion. In *The Great Apes* (D. A. Hamburg and E. R. McCown, eds.), Benjamin/Cummings, Menlo Park, CA, pp. 481–489.

Wrangham, R. W. (1993). The evolution of sexuality in chimpanzees and bonobos. *Human Nature* 4: 47–79.

Wrangham, R. W. (1999). Evolution of coalitionary killing. *Yearbook of Physical Anthropology* 42: 1–30.

Wrangham, R. W. (2000). Why are male chimpanzees more gregarious than mothers? A scramble competition hypothesis. In *Primate Males: Causes and Consequences of Variation in Group Composition* (P. M. Kappeler, ed.), Cambridge University Press, Cambridge, pp. 248–258.

Wrangham, R. W. (2002). The cost of sexual attraction: Is there a trade-off in female *Pan* between sex appeal and received coercion? In *Behavioural Diversity in Chimpanzees and Bonobos* (C. Boesch, G. Hohmann, and L. F. Marchant, L.F., eds.), Cambridge University Press, Cambridge, pp. 204–215.

Wrangham, R. W. (2008). Why the link between long-term research and conservation is a case worth making. In *Science and Conservation in African Forests* (R. Wrangham and E. Ross, eds.), Cambridge University Press, Cambridge, pp. 1–8.

Wrangham, R. W. and Bergmann Riss, E. vZ. (1990). Rates of predation on mammals by Gombe chimpanzees, 1972–1975. *Primates* 31: 157–170.

Wrangham, R. W. and Nishida, T. (1983). *Aspilia* spp. leaves: A puzzle in the feeding behavior of wild chimpanzees. *Primates* 24: 276–282.

Wrangham, R. W. and Ross, E., eds. (2008). *Science and Conservation in African Forests: The Benefits of Longterm Research.* Cambridge University Press, Cambridge.

Wrangham, R. W. and Smuts, B. B. (1980). Sex differences in the behavioural ecology of chimpanzees in the Gombe National Park, Tanzania. *Journal of Reproduction and Fertility, Supplement* 28: 13–31.

Wrangham, R. W., Clark (Arcadi), A. P., and Isabirye-Basuta, G. (1992). Female social relationships and social organization of Kibale Forest chimpanzees. In *Topics in Primatology, Vol. 1: Human Origins* (T. Nishida, W. C. McGrew, P. Marler, M. Pickford, and F. B. M. de Waal, eds.), University of Tokyo Press, Tokyo, pp. 81–98.

Wrangham, R. W., McGrew, W. C., de Waal, F. B., and Heltne, P. G., eds. (1994). *Chimpanzee Cultures.* Harvard University Press, Cambridge, MA.

Wrangham, R. W., Wilson, M. L., and Muller, M. N. (2006). Comparative rates of violence in chimpanzees and humans. *Primates* 47: 14–26.

Wrangham, R. W., Koops, K., Machanda, Z. P., Worthington, S., Bernard, A. B., Brazeau, N. F., Donovan, R., Rosen, J., Wilke, C., Otali, E., and Muller, M. N. (2016). Distribution of a chimpanzee social custom is explained by matrilineal relationship rather than conformity. *Current Biology* 26: 3033–3037.

Wroblewski, E. E., Murray, C. M., Keele, B. F., Schumacher-Stankey, J. C., Hahn, B. H., and Pusey, A. E. (2009). Male dominance rank and reproductive success in chimpanzees, *Pan troglodytes schweinfurthii. Animal Behaviour* 77: 873–885.

Zamma, K. (2002). Leaf-grooming by a wild chimpanzee in Mahale. *Primates* 43: 87–90.

Zamma, K. (2005). Rejecting a bit of meat to get more. *Pan Africa News* 12: 8–10.

Zamma, K. (2014). What makes wild chimpanzees wake up at night? *Primates* 55: 51–57.

Index